DARK SIDE OF THE MOON

WERNHER VON BRAUN, THE THIRD REICH, AND THE SPACE RACE

WAYNE BIDDLE

W. W. NORTON & COMPANY

NEW YORK · LONDON

For information about permission to reproduce selections from this book,
write to Permissions, W. W. Norton & Company, Inc.,
500 Fifth Avenue, New York, NY 10110

For information about special discounts for bulk purchases, please contact
W. W. Norton Special Sales at specialsales@wwnorton.com or 800-233-4830

Manufacturing by The Courier Companies, Inc.
Book design by Chris Welch
Production manager: Andrew Marasia

Library of Congress Cataloging-in-Publication Data

Biddle, Wayne.
Dark side of the moon : Wernher von Braun, the Third Reich, and the
space race / Wayne Biddle. — 1st ed.
p. cm.
Includes bibliographical references and index.
ISBN 978-0-393-05910-6 (hardcover)
1. Von Braun, Wernher, 1912–1977. 2. Rocketry—Germany—Biography.
3. Rocketry—United States—Biography. 4. World War, 1939–1945—Science.
5. Space race—United States—History—20th century.
6. Germany—Politics and government—1933–1945. 7. Cold War. I. Title.
TL781.85.V6B53 2009
629.4092—dc22
[B]
2009015572

W. W. Norton & Company, Inc.
500 Fifth Avenue, New York, N.Y. 10110
www.wwnorton.com

W. W. Norton & Company Ltd.
Castle House, 75/76 Wells Street, London W1T 3QT

1 2 3 4 5 6 7 8 9 0

DARK
SIDE
OF THE
MOON

FOR JEAN

CONTENTS

Introduction ix

1 A JUNKER'S LIFE 1

2 MEMORIES OF DEFEAT 17

3 "HIGHLY TECHNOLOGICAL ROMANTICISM" 33

4 AN HEIR OF CREDIBILITY 46

5 CHILDHOOD'S END 56

6 "FINGERS IN THE PIE" 70

7 SUPREME ZEAL 84

8 GRAND AND HORRIBLY WRONG 96

9 DEPRAVITY 109

10 "A PSYCHOLOGICAL BLOCK" 127

Epilogue 145

Notes 153

Selected Bibliography 199

Photograph Credits 205

Index 207

INTRODUCTION

The saying that "no one is voluntarily wicked nor
involuntarily happy" seems to be partly false and partly true;
for no one is involuntarily happy, but wickedness is voluntary.
—*Nicomachean Ethics*

THE SUBJECT OF moral responsibility has a long history, stretching at least as far back as the fourth century BC. In Book Three of his *Nicomachean Ethics*, Aristotle discussed how to decide whether someone should be praised or blamed—either way, that is—for an action. Simply put, a person is morally responsible for something if he was causally responsible for it, if he was in a position to know that it would come about, if what he did was freely undertaken, and if he was fully rational. As courts of law demonstrate every day, real life is infinitely complicated and each segment of this statement has been masticated by philosophers ever since, but it is fair to say that there is broad agreement in Western culture that if all four conditions are met, then an individual is on the spot for better or worse.[1]

The issue of whether scientists are responsible for the outcomes of their work has been especially turbid since they became powerful enough during the last century to alter the landscape. In 1962, the chemist and sociologist Michael Polanyi (1891–1976) argued in an influential essay called "The Republic of Science" that the practical results of pure science are often unforeseen; therefore a scientist may not be liable for what others do with them. He recalled that in April 1945, he and Bertrand Russell were asked on a British radio show what applications might arise from Einstein's $E = mc^2$, and they both drew a blank.

Laws of nature are morally neutral, everyone agrees. But so-called pure science—the discovery or extension of fundamental theory—is practiced by only a small fraction of the profession. The rest take accepted principles and apply them to carefully defined problems with definite notions about the solutions, which may or may not have practical outcomes. Most engineers, of course, are directly involved with applications in the everyday world. Nonetheless, the separation between scientists and engineers is often impossible to delineate. Yet the convention that scientists as a group exist in a stratum detached from and untainted by common sociopolitical forces, and are thus somehow above reproach, has proved remarkably durable, albeit weakened by the past half-century of disasters clearly traceable to their activity.

"Do you think scientists should be blamed for wars? Einstein? He looked for fundamental truths and his formula was used for an atomic bomb. Alexander Graham Bell? Military orders that kill thousands are transmitted over his telephone. Why not blame the bus driver who takes war workers to their factories? How about movie actors who sing for the U.S.O.?" These comments may sound philistine today, but they were made six years after the end of World War II in a respectful article in *The New Yorker* magazine by a man then well on his way to becoming one of the era's most famous and revered technologists. The German "rocket scientist" Wernher von Braun, who had played a principal role in creating the V-2 "wonder weapon" for the Third Reich, had come to America under U.S. Army auspices to continue his work. What he had really been doing all along was developing the means to travel into outer space, he claimed time and again. Most of his audience never doubted it.

Later generations largely forgot about him, as the imperatives of the Space Age grew quaint. But it is reasonable to posit that no other public figure of the twentieth century was forgiven so much as Wernher von Braun, so that he be allowed to pursue his dream. The army gave him a fresh identity by classifying for decades the most malevolent details of his pre-1945 life. He needed little else besides more of the incredible good luck that had propelled him out of Hitler's Germany into General Eisenhower's America.

There has been much biographical writing about von Braun over the years. Because of government secrecy and popular disinclination, most of it was uncritical until long after he died (of cancer) in 1977. As the archives opened up and cold war restrictions on traveling in eastern Germany relaxed, a few journalists and historians performed the investigative toil of straightening out a record that had been warped by public relations men and pervasive sycophancy.

My own entry into this fray came about because I was one of innumerable members of the postwar baby boom generation who enjoyed von Braun as an inspirational narrator on Walt Disney's television shows about space travel during the mid-1950s. Young boys, especially, built models and read science fiction and believed that "mankind" would explore the solar system in our lifetime. It is ironic that von Braun had a similar boyhood in the 1920s and managed to transfer it directly into our brains. Also like many of my age, the political atmosphere ten years later led me to wonder skeptically how a Nazi weapons builder could become an American hero. Yes, he played a central role in sending astronauts to the moon, but, by contrast, LBJ would never be forgiven for Vietnam, so why was this man let off the hook? Or so the thinking went.

The question was not merely an artifact of a credulous childhood, but a window into a core phenomenon of American technology and culture. Von Braun moved so seamlessly from Peenemünde, Pomerania, to Huntsville, Alabama, because millions of people wanted him to, because the secrecy and some of the obsessions of the Third Reich were not entirely different from those of postwar America. They wanted to believe in his prophecies, his genius, and his goodness, no matter what. This is why his life remained of interest to me, not because of the exciting hardware he produced, which like all hardware is ultimately trivial when divorced from its social and political context.

My goal for this book was to bring von Braun truthfully through the first thirty-three years of his life in Germany to America in 1945 and then tell how he blended into the late-1940s and 1950s. That was when the torquing of his persona took place, which I believe to be one of the central conundrums of that weird time of anti-Communist hysteria and progress worship. After *Sputnik*, I let his story go, because then he became all about hardware, so to speak. Other writers have brought him almost day-by-day through Project Apollo into the 1970s, but I find that the story gets fetishistic then and of interest mostly to buffs.

After all these years, historians still tend to divide into two factions: those who cheer von Braun as a space pioneer and those who condemn him as a Nazi. The fact that he was both has not brought the two sides together, but it is probably impossible to both cheer and condemn in this case and not sound stupid. He was able to thrive and his success needs to be explained. An article I wrote about this for the op-ed page of *The New York Times* in 1992—when officials of the German government and aerospace industry were planning to celebrate the fiftieth anniversary of the V-2's first successful launch (they changed their minds)—was illustrated with a cartoon of von Braun in jack boots and swastika armband, his arm raised in a Nazi salute that ended in an Apollo capsule heading for the moon. Op-ed contributors are not consulted about illustrations, and as soon as I saw it in the paper, I was dismayed; I knew that, as a journalist, some people who were part of his life would now never talk to me. Other people who were also part of his life, but who had been ignored or afraid to talk ever since leaving the slave camp called Dora where V-2s were constructed, made themselves available. The cartoon was dead-on, actually, and I have come to be fond of it.

In April 1995, I attended the fiftieth anniversary celebration of Dora's liberation in Nordhausen, Germany. When I first saw the camp and the underground tunnels of the V-2 factory that it had supplied with forced laborers (some 20,000 of whom died in the process), the obvious fact that they were a single entity, inseparable one from the other, froze me in my tracks in more ways than one. The shock of grasping the size of the distortion perpetrated by von Braun and his American apologists, who either covered up the place's horrors or maintained that being involved with the V-2 was different from being involved with the slaves—that science was above society, in largest terms—was so strong that trying to convey the nature of the falsehood suddenly seemed futile. Who would believe this without seeing it? Germans might cover it up to save their necks, but why would Americans help them?

One falls back on Aristotle. But the way out of this bind ultimately came from the people most damaged by the camp. On that raw gray Thuringian spring day, as the old French Resistance fighters—who had been imprisoned by the SS at Dora, worked in the brutal tunnels, and somehow survived—climbed down the steps of a tour bus that had brought them back to the camp's drab entrance for commemoration ceremonies, children of the current citizens of Nordhausen silently handed each of them a white rosebud. I shall never forget the men's faces as they struggled to appreciate the gift. They all held onto the roses. There is something going on here that is bigger than flying to the moon, I thought to myself.

ACKNOWLEDGMENTS

I wish to gratefully acknowledge the help of the following organizations and individuals:

The Smithsonian Institution's A. Verville Fellowship at the National Air and Space Museum provided invaluable access to the museum's library and archives, as well as collegial exchange with curatorial staff, including Gregg Herken, Tom Crouch, and Michael Neufeld. I am indebted to the staff of the KZ-Gedenkstätte Mittelbau-Dora museum and research center for assistance on numerous occasions, as well as to the French Association des Déportés de Dora, Ellrich, Harzungen et K° and its American affiliate. Of the many former Dora prisoners who patiently discussed their experience, I am especially grateful to Yves Béon, Guido Zembsch-Schreve, Jean Mialet, Georges Jouanin, Georges Soubirous, Louis Garnier, Andre Marais, Lucien Colonel, G. P. Michajluk, and Georgi Loik. I also wish to thank Gretchen Schafft, Tom Gehrels, and Susan Bardgett for their cordial regard.

The staffs of the Imperial War Museum in London, the Deutsches Museum in Munich, the Library of Congress and the George Washington University Library in Washington, DC, the Auburn University Library in Alabama, the National Archives in College Park, Maryland, and The

Johns Hopkins University Library in Baltimore provided gracious assistance with collections.

In 2001, the Technical University in Berlin invited me to teach about the technology and culture of the "space race" as a visiting professor; I wish to thank professors Karl Heinz Stahl and Clemens Schwendler for their generosity. On two research trips to Peenemünde, in 1995 and 2001, the Historisch-technisches Informationszentrum there provided orientation to the Usedom region's points of interest. I am especially grateful to Clemens Schwendler and Christian Ludwig for their perseverance in helping to find the ruins of *Prüfstand VII*.

I wish to express singular gratitude to Gretel Furner for her friendship and acumen as a teacher and translator of the German language.

Jean McGarry both cast her sharp writer's eye upon every word and kept the home of two working authors in good order, a task that transcends appreciation.

I must finally thank several individuals for their professional and personal kindness: Angela Von der Lippe, Erica Stern, Julie Tesser, Zoe Pagnamenta, Kai Bird, Peter Carey, Mitchell Levitas, Lee Atwood, John Pike, Arno J. Mayer, Thomas P. Hughes, Paul and Mary Medlicott, Liz Bird and Brian McInally, and Mimi Harrison (who suggested this subject long ago).

Wayne Biddle
Baltimore, February 2009

The only historian capable of fanning the spark of hope in the past is the one who is firmly convinced that *even the dead* will not be safe from the enemy if he is victorious. And this enemy has never ceased to be victorious.

—Walter Benjamin, *On the Concept of History*

DARK
SIDE
OF THE
MOON

GERMANY AND THE THIRD REICH

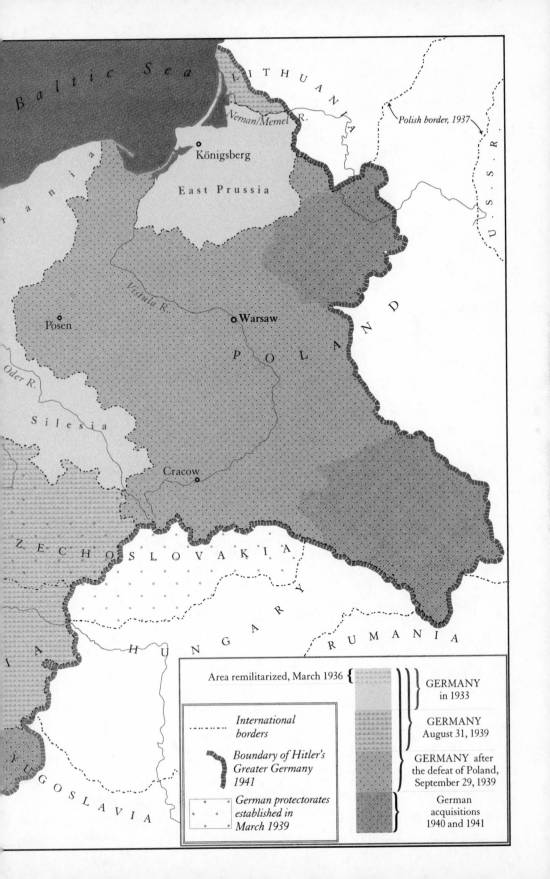

Baltic Sea

LITHUANIA

Neman/Memel R.

Polish border, 1937

Königsberg

East Prussia

U.S.S.R.

Vistula R.

Posen

Warsaw

P O L A N D

Oder R.

Silesia

Cracow

C Z E C H O S L O V A K I A

H U N G A R Y

R U M A N I A

Y U G O S L A V I A

Area remilitarized, March 1936 {

GERMANY
in 1933

International
borders

GERMANY
August 31, 1939

Boundary of Hitler's
Greater Germany
1941

GERMANY after
the defeat of Poland,
September 29, 1939

German protectorates
established in
March 1939

German
acquisitions
1940 and 1941

1

A JUNKER'S LIFE

THERE IS A vexatious delight, well known to children but sometimes to adults with certain technical predilections, that involves launching a missile through the air. The projectile might be a stone pitched with no more force than a young arm can muster on a summer's day, or a 30-story-tall colossus with multi-million-horsepower engines that was once beyond imagination. Let it go, watch it rise, see what it hits. If it can be ridden somehow, by a mouse or a man, so much the better.

No matter how it looks at first glance, this is not a childish pursuit. National treasure and many lives have been spent on its most spectacular forms, which require long concentration of collective will. Like many ambitions, what seems innocent to some appears evil to others. Yet at times, particularly during war, there should be no confusion. The rockets are meant to kill.

This act of adventure or aggression has proved more than once to strike a rapturous chord in romantic instinct, as though Henry Adams's Dynamo and Virgin were conjoined, carrying their worshippers to heaven as poetically as blasting them to hell. Escapist fantasies erode technical discipline and soon the greatest machines' amorality, their thunderous autonomy once off the ground, becomes part of the price of scientific imagination outside the reach of value judgments. *Rocket scientist*—the term itself resonates with both genius and malevolence. In the name of twentieth-century progress there was much of each, but rarely were they condoned with such simultaneousness, as though a boy could build glorious towers with his toy blocks and torture helpless animals with them, too, all the while encouraged by proud parents.

Connections abound to childhood's dreamscape, as inescapable as any nightmare of mass destruction. It is usually this way with demons that possess a beautiful face. Grown men will make pacts to consort with them, relying on lies to hide their sins if they survive.

THE SPRING OF 1933 brought cold rain as always across northern Europe, from the Meuse River in the west to the Vistula in the east. These water-ways were the ancient boundaries of empires that had shifted so many times that the names of imperial cities and feudal villages alike comprised a history of French, Dutch, German, and Slavic fortunes. Though some places—Prussia, Posen, Silesia—had been so carved up by war that they no longer existed except in prideful memory, the region's grain fields still grew as lush as ever with the iridescent green of new growth. After ample rain there would be mountains of wheat and rye and three cuttings of timothy or alfalfa, enough to feed the people and as much of their live-stock as anyone cared to bring to market in Berlin. This was the rhythm of the ages, the reassuring cycle of soil and seed and patient husbandry that for centuries had been enough to define the scope of human dignity. From here, the Great Depression that started far away in New York looked like a suitable curse on greedy bankers, those "unscrupulous money lend-ers" as the new American president himself called them.[1] To be left unbothered amid such beauty was what almost any young man or woman might wish for in a lifetime, setting a utopian standard for contentment even among cosmopolitans who could hardly tell a Holstein from a Lipizzaner.[2]

A measure of stability also grew here from living among one's own kind, from a *Zusammengehorigkeitsgefuhl*, or "feeling of belonging-togetherness," defined in myriad provincial ways but nonetheless apparent to all.[3] Never mind that the mixing of kinds was deeper than many cared to admit. Home should be a repository for familiarities, a haven from strangeness. Since the time of Charlemagne, Europe had divided endlessly into a discordant mosaic of homelands whose precise boundaries shifted according to the ambitions of princes or merchants but whose masses of ordinary people cherished nothing more than the equilibrium of their life at home. That the Great War had fomented social and technological forces that were antithetical to such continuity was something most unreflective folk could comprehend only when it reached their doorstep. Their instinct then was to protect the hearth, to preserve the customs, to glorify one's own kind, to eliminate the intruder. The last impulse would grow murderous against people who had been kept separate for centuries by state edicts, by religious laws forbidding assimilation, and by sheer bloodymindedness. Inexorably, belonging together turned sinister, far beyond nationalism and *Volksgemeinschaft*.[4]

Even though Europe had barely recovered from one catastrophe that had eviscerated an entire generation, new killers of great category seized control of events in the spring of 1933, unleashing another disaster that would make the first seem quaint. Only twenty years lay between Germany's ruination in 1919 and its re-emergence as a belligerent world power in 1939. This time the tragedy would be so vast that there would be no escape for anyone, not even the unborn. It would spread through the rest of the century like brimstone that refused to cool, igniting personalities that might otherwise have remained obscure while smothering many that could thrive only in less brutal times.

On March 23, 1933, the middle son of a Prussian family of noble title reached his twenty-first birthday. Wernher Magnus Maximilian Freiherr von Braun, grandson of an Imperial officer descended from a thirteenth-century knight, was situated by birth to receive all the resplendent gifts his homeland could bestow.[5] The spelling of his first name, with an *h*, was a linguistic echo of the Middle Ages, a family affectation that signaled residence in a class apart. The inherited *Freiherr*—meaning baron, but literally lord or ruler, another remnant from the feudal past—still carried weight in a society that had maintained an absolute monarch longer than any other in the West. With a blonde, crystalline blue-eyed visage and manners that somehow reminded girls in his Berlin student coterie of Oscar Wilde's notorious lover, Lord Alfred Douglas, he stood out as a slightly anachronistic rich boy, perhaps, but little more.[6]

To wit, he spoke French and could play Beethoven's "Moonlight Sonata" by heart—dilettantish achievements, but reasonable insulation against loutishness.[7] One might say, without too much Romantic hyperbole, that the seven hundred years of German civilization since Ritter Henimanus de Bruno sowed his seed across Silesia had all devolved upon the head of that knight's distant kin, this modern baron. It being 1933 and not 1833, some might also say that this was *Quatsch* (rubbish) except that many Germans still believed they were God's chosen people, that men could be socially advanced and loyal to their heritage at the same time.

Much more than a family dialectic was being fulfilled that spring, however, and young von Braun was no more likely to extract himself from the ongoing cataclysm of German politics than from his genealogy. No intelligent German in the 1930s, young or old, could be apolitical. In a society so polarized, there was no neutral ground except obliviousness, and Wernher lived closer to the fray than most. His father, Magnus Alexander Maximilian Freiherr von Braun, had just completed a role as stirrup-holder in Hitler's elevation to power, finding his own political career trampled in the process, like so many others'. As the elder baron represented the unraveling of a class that not much earlier had seemed unshakable, his son, already a weapons researcher for the German Army, stood on the brink of a new order that would turn traditional hierarchies—along with everything else in German life—to its hellish purpose.

IF A SINGLE springtime could be erased from modern history, so that what germinated then were nipped in the bud, few would argue that it should not be this one. When Hitler ascended the steps of the presidential palace in Berlin on the icy morning of January 30, two months before Wernher's birthday, parliamentary government in Germany ceased. Hitler had come to accept from octogenarian President Paul von Hindenburg, the dotterel *Ersatzkaiser*, a formal invitation to be Reich chancellor—the fourth man in less than a year to hold that battered office. When the fascist opponent of democracy joined hands with the conservative opponent of change, a fusion of dissimilar political metals took place that would electrocute the world. That night, from seven o'clock until after midnight, 25,000 of Hitler's paramilitary followers staged a torchlight parade through the Brandenburg Tor and along the Wilhelmstrasse, claiming for themselves the capital's ancient path of victory as they filled the city with "insensate tumult."[8] Hindenburg, illuminated by police spotlights and only vaguely aware of what was happening, took their salutes from a window of his official residence, sanctifying the transference of power. On the next day, the National Socialist German Students' League paraded to the stock exchange and screamed "Judah perish!" at the brokers.[9]

The National Socialists had made their breakthrough from *Saalschlacten* (brawlers) to mainstream as a petty bourgeois and agrarian protest movement, aided by ultranationalist and conservative establishment circles and antiunion business leaders.[10] Hitler garnered mass appeal as he nursed German confidence, which had been shattered by the Great War and subsequent economic chaos. Foreigners who read about Hitler's appointment in newspapers or watched his handshake with Hindenburg in cinema newsreels, without seeing for themselves the mayhem in Berlin's streets, perceived a constitutional transfer of power and were usually satisfied that these uncouth Nazis were not revolutionaries *à la guillotine,* but champions of order and stability. In America, after all, Franklin D. Roosevelt was telling his own dispirited people that "the money changers have fled from their high seats in the temple of our civilization," warning that if necessary he would seek from Congress executive power as broad as that available in case of foreign invasion to deal with the economic disaster. "We may now restore that temple to the ancient truths."[11]

Of course, the German ruling class, which like all such strata consisted of men who thought of themselves as cultivated and humane, knew very well what was happening in the streets.[12] There were not just long lines of forlorn unemployed, as in other cities of the industrialized world. The first months after the Nazis came to power were widely known as the *Blutkrieg,* or Blood War, with grisly brawls between Hitler's thugs and opposition gangs, especially Communists. The elites might have regarded them all as philistines, but social fancies and imperial experience had created a value system that could only be reconciled with ideals of justice by not scrutinizing life too closely. In this way, the right-wing establishment tolerated Hitler far longer than hindsight can comprehend.

Four weeks after winning the chancellorship came the real coup d'état, when an unemployed Dutch anarchist set the Reichstag afire on the night of February 27. Like the accelerating nightmare of German politics, the great edifice burned fast, with the monumental inscription "Dem Deutschen Volk" high across its entrance blackened by flames. Hitler lost no time issuing a decree "for the Protection of the People and the State" against a fantasized Bolshevik takeover. Constitutional civil liberties—speech, assembly, press, privacy—were quashed the next morning, followed by a wave of arrests and murders of suspected Communists and Social Democrats that grew to Terror proportions. Storm troopers ruled the cities.

Early in March, anti-Jewish laws were introduced legalizing the purge of the civil service and judiciary, of universities and medicine, and stripping political and racial undesirables of their citizenship.[13] What had always been between the lines was now in bold print. The great German film conglomerate UFA promptly dismissed all its Jewish workers. In Munich on March 20, Heinrich Himmler, the leader of the SS (from *Schutzstaffeln*, meaning defense or protective squadron), announced the opening of the first "concentration camp for political prisoners"—Communists and Social Democrats—at Dachau.[14]

On March 23, Hitler presented the Reichstag, temporarily convened in the Kroll Opera House, with a bill through which he could rule by dictatorial edict. The opera house was cordoned off by armed SS units, originally organized as his personal guard but now a cache for pure fanaticism. Fear that dissenters would be beaten up or jailed hung thickly over the session, like the swastika flag of theatrical dimensions suspended at the back of the stage. Hitler, wearing a brown shirt after weeks of dressing in civilian clothes, threatened the political parties that the consequence of any opposition would be civil war.[15] Though the Reichstag had a well-deserved reputation as one of the most contentious political bodies in the world, few cared to test his rhetoric on this particular day.

With many of his opponents already incarcerated or dead, and after only the Socialist minority voted against him, the congress was dissolved until it pleased the führer to summon it again.[16] Now there was no need for a civilized veneer. Throngs of Nazis swarmed forward, arms raised in their catatonic salute, singing the "Horst Wessel" anthem: *Die Fahne hoch! Die Reihen fest geschlossen!*[17] The Weimar Republic collapsed—felled by disunity, ambivalence, and horrendous luck. A political genius whose primary motivating force was hate now held absolute power over a society renowned for intellect and order, a bizarre contradiction that he was uniquely ready to exploit.[18] "You don't go railing against the ocean," mourned one prominent liberal about the extinction of anti-Hitler voices from left to right.[19] The first boycott of Jewish stores occurred on March 28, affecting thousands of shops nationwide and bringing whole sections of many towns to a standstill. By summer, a one-party state was in full bloom, with the army, trade unions, churches, paramilitary veterans' brigades, and rival political groups all neutralized as potential opponents.

The Kroll Opera drama, headlined across the front page of every major newspaper in the world, thus happened to coincide with Wernher's twenty-first birthday, an auspicious juncture by any measure. There is no record of what he did or thought about on that day—after a whole lifetime he would leave behind only a few oblique words about the Third Reich—but his family was anything but detached from these events. It is fair to assume that the occasion was sharply amplified by the fresh wreckage of his father's career in government. The elder von Braun, paragon of East Prussian estate owners and Wilhelminian conservatives, had just turned fifty-five in February and would spend the rest of his long life contemplating the catastrophe he had been party to. In this simple regard he was infinitely more fortunate than millions of his countrymen destined to perish during the next dozen years, yet still a *mutillé de la guerre* compared to his son, whom astonishing fortune would eventually propel from vanquished to victor as though there were nothing unfriendly in between.

IIIII

Born in the bucolic village of Neucken, East Prussia, in 1878, Magnus von Braun lived until the age of eighteen on a 600-acre estate near Königsberg (now Kaliningrad, in the Russian exclave between Poland and Lithuania) that he always referred to simply as Heimat, home.[20] His use of this term reflected its status as one of the German language's keywords, carrying conservative political implications in much the same way as *Volk, Nation,* and *Staat.*[21] To talk of one's Heimat signaled a sense of identity with the land, a belief in the influence of nature on national character. It was the opposite of *kosmopolitisch,* the codeword for disparaging city-dwelling Jews. In the late-nineteenth century, various cultural and historical organizations devoted to nostalgic visions of the German past constituted a Heimat movement that for the most part was benignly provincial but could be tinged with racism, anti-Semitism, and prejudice against urban society.

Acquired by his great-grandfather in 1804 near the end of a period of land speculation in the region, the tract was one of many east of the Elbe River originally ruled during the twelfth and thirteenth centuries by the Teutonic Knights, a bellicose religious order founded by former Crusaders. After conquering an indigenous pagan tribe, known as the Pruzzen, with the aid of the same contagious microbes that had decimated their own ranks in the Holy Land, the Knights had gradually parceled out their lands to noblemen and others who agreed to serve in the fraternity's military adventures. Those who resisted, like the Estonians, were routed with crossbows, weapons of such infamy that Conrad III had declared them illegal in his German kingdom.[22] By modern times, after their violence had long been romanticized, any connection, real or imagined, to these tightly organized warlords was coveted by nationalistic Germans. To have been born and raised on their old turf was tantamount to carrying forth their genes.

An expansive "new" house—which Magnus described in classic Heimat metaphor as a "simple folk melody that had become stone"—was completed in 1806. Surrounded by towering oaks and boxwood hedges trimmed symmetrically in the French style, the house was situated near a man-made lake for boating and swimming. An old seven-stemmed Linde, or limewood tree, Germanic symbol of domestic tranquility, provided shade for children who grew up "in the fullness of the moment, although we lived, without knowing it, in the past," he would recall many years later.[23] He also vividly remembered an inscription inside a cupboard dated 1756, the beginning of the Seven Years' War against Austria, France, Russia, and Sweden: "Learn to live economically with what you have, because these are hard times."[24] This was sage advice for any epoch in a region where the landed gentry, whose primary role was to fill the Prussian officer corps, often lacked either business or agricultural expertise.

Typical of this milieu, which equated itself with the glories of German history, the family kept icons of high culture on prominent display. Among the treasures conserved in vitrines, which Magnus remembered as filling him with "respect and dread," was a silver sugar spoon with fishbone stem, a wedding gift from Königsberg neighbor Immanuel Kant.[25] There was also a gold snuff box which Czar Alexander I had given in gratitude to his great-uncle Wilhelm von Braun, a riding master and ordnance officer for the Auer dragoons during the years of alliance with Russia against Napoleon. Cannonballs from the 1807 battle of Preussisch-Eylau, when a desperate remnant of the Prussian Army had briefly turned back Napoleon's advance, were imbedded in the stone walls of a house later built by Magnus's father after marriage to Eleonore von Gostkowsky, whose father had fought there. In this country a man gave his bride munitions, not roses, to prove his devotion.

Bloodlines were thus entwined with battle lines, securing the cohesive devotion of generations to an elitist system of hereditary power. The Prussian Army made Prussia in the seventeenth and eighteenth centuries, and then transmitted its traditions to the German Reich in the nineteenth and twentieth. Military virtues such as rectitude and obedience, taken to an extreme that often seemed grotesque to foreigners, formed the armature of these ways. The ideal Prussian officer might exercise individual judgment, but he was never encouraged to speculate. As a contented adjunct of the conservative oligarchy, his actions in the wider political sphere were rigidly conditioned. As such, he prospered and served as a role model for young men who might never experience a cavalry charge.

Like many other Prussian estates, the von Braun home was pillaged and occupied by the French during the height of Napoleon's rule, from 1807 to 1812. The family's household possessions were stolen or vandalized while the farm supplied the French Army. A diary recorded the indignities suffered by Magnus's great aunt, who was reduced to sipping her tea from a cook pot after Frenchmen "who behaved worse than Russians" smashed her china. Saved from total ruin by a sympathetic French duke whose class loyalties evidently transcended nationalism, the estate was restored after Napoleon's defeat in 1814 and eventually passed into the hands of Magnus's father, Imperial Prussian Lieutenant Colonel Maximilian Freiherr von Braun.

The long-term effect of this marshal history was to set in stone the political reliability of Prussian traditions, euphemistically called *Gutsherrschaft* (literally "rule over an estate"), that maintained an abyss between aristocracy and commoners.[26] Not surprisingly, these Prussian *Junkers* (from *junc-herre*, young noblemen) became known for a narrowness that turned easily into bigotry—the line between pride and arrogance being invisibly thin.[27] The human species, it was said, began with the lieutenant.[28] As an adult, Magnus von Braun made much of the fact that his favorite childhood playmates had been the coachman's sons, as though evidence of how class barriers could be happily bridged. But in a region of sprawling estates, a servant's children were likely to be the only ones for miles around, merely emphasizing the abyss when the time came to be schooled or married. The Prussian officer corps had tried to embrace the middle class only after the debacle of defeat by Napoleon, when it became clear that the Prussian people were unhitching themselves from the fate of their Junker-dominated government and army, which plainly regarded them as cannon fodder.[29]

Unlike the coachman's sons, who were illiterate, Magnus von Braun was tutored at home until the age of twelve, and then sent to the exclusive *Wilhelmgymnasium* in Königsberg for a classical education. Despite growing up with the famous philosopher's silver spoon, he never developed much affinity for Kant, but admired Goethe from the start.[30] From Königsberg it was a natural step to the University of Göttingen to study law, where there were certainly no servants' children in his circle. He lasted only a year at rigorous Göttingen, however, returning to Königsberg to finish his studies at the less prestigious university there.

After graduation in 1899, with two older brothers ahead of him in the line of inheritance, he stayed in Königsberg to train for the civil service—a traditional Junker prerogative—and then took various minor posts around the region. He was of a class that needed close ties to the government for social and economic sustenance, having few riches of its own beyond possession of land. In 1907 he received a commission from the Prussian Interior Ministry to study the British banking system in London. Britain had just tipped the world's military scales by launching the HMS *Dreadnought* and joining a triple entente with France and Russia to counter the Triple Alliance of Germany, Austria-Hungary, and Italy. There was no shortage of German expertise on international banking, but it was clearly an opportune moment for an ambitious young *Burokrat*—whose career was still tepid—to spend a year across the Channel.

Living in a Kensington Garden boarding house, he took special notice of two Jewish women there, finding it remarkable that he never witnessed discrimination against Jews in England. The English Jew, he marveled, was treated "through and through like an Englishman" and was "far less international than Jews in other countries."[31] With another boarding house resident, a young man from India, he held long palavers about Hinduism and the caste system. Evidently finding parallels with his native Prussia, he grew fearful about what would happen if class barriers ever broke down. "After all, people are no more than good-hearted sheep [*gutherzige Hammel*]," he wrote, "with the instincts of predatory animals [*mit Raubtierinstinkten*], prepared to follow any shepherd who has a stick and watch-dog, but also prepared to kill if this shepherd commands it."[32]

The young Junker especially admired the workaday concept of the English gentleman, how City banks often dealt with their customers on the basis of a gentleman's word rather than by formal contract. To him, being a gentleman meant "reliability, reserve, self-assuredness without

snobbishness, and rejection of ordinary mass-instincts."[33] These were per-
mutations of classic Prussian military ideals, yet also typical sentiments of
privileged men throughout the Western world on the eve of a war that
would smash old illusions about who was best born to lead.

The British Empire, with its world-embracing power and wealth, was
still something to emulate. When Magnus went home to Berlin in 1908,
he returned to another empire at its apogee. The traffic was nothing compared to Oxford Street, and Germany had failed to find adequate outlets
beyond Europe for its ambitions, but it was the acknowledged heart of the
Continent. The kaiser's autocratic court, which spent summer in Potsdam
and winter in Berlin, set the social tone as indelibly as that of the sybaritic
Edward VII. Like generation after generation of his forebears, Magnus's
political compass was set by unquestioning loyalty to the kaiser, distaste
for mass movements, and suspicion of democracy. Were cultures born like
animal organisms to live and die, he wondered, or did civilizations rise
and fall with the creative powers of their elites? These two outlooks, the
former German and the latter English in his mind, each in its own way
prone to justify conflict, were the taproots of his personal politics as he
entered his thirtieth year, still unmarried and childless.[34]

As Junkers, the von Brauns would be expected to dwell within privileged circles of government and conservative wealth. The world of music,
art, and high culture would naturally be claimed by them, but regarded
with suspicion and contempt at its unorthodox frontiers. The family did
not count any scholars, writers, or artists among themselves, nor anyone
with an amateur or professional interest in science and technology. This
last category would be especially apropos in retrospect, but it was empty.

Science was very much in the air, however, in Germany perhaps more
than in any other country. The two decades before World War I witnessed
a radical transformation of science that would revolutionize many fields of
knowledge.[35] Germans were leaders in this epoch when the natural sciences reached their richest bloom in Germany, shattering principles that
had been considered absolute truth. Wilhelm Roentgen discovered X-rays
in 1895, leading through studies of radioactivity to the obliteration of a
basic tenet of chemistry, the immutability of elements. Max Planck pub-

lished his quantum theory in 1900, attacking the fundamental assumption that nature's forces are continuous. And Albert Einstein presented his special theory of relativity in 1905, shaking classic concepts of space and time. Suddenly, nature could no longer be simply sensed, described only in abstract terms. Though comprehended by few people outside scientific coteries, this new knowledge reverberated in the social sciences, the humanities, and beyond into literate society, creating anxiety about the hypothetical character of many cherished ideas. At the same time, Germany was the world leader in science-based industries such as dyes, pharmaceuticals, electricity, and agriculture, producing unprecedented prosperity with the support of government-funded research. The older elites might be well-insulated from the social change that accompanied such ferment, but their more inquisitive children would be immersed in it for better or worse.

True to his milieu, the first turning point in Magnus von Braun's career occurred at a dinner party, where he was noticed by an undersecretary from the Ministry of Trade. Several days later he read in a newspaper that he had been appointed an assistant to Clemens von Delbruck, the trade minister.[36] This was how the autocracy maintained itself, choosing new members by their appearance at the right social occasion and assuming the rest would fall into place. Magnus's social skills were evidently more useful than his research acumen, because he soon settled in as a sort of personal secretary for Delbruck, managing his daily agenda of meetings and receptions, thereby gaining an insider's view of "the foxhole known as government." He was also expected to help keep Delbruck up to date on current affairs, but was scolded by his master for attempting to lead him on rather than simply inform.

In these prewar years, the trade minister's palace on the Leipziger Platz in Berlin became a showplace for modern German style and its commercial accoutrements.[37] At official dinners for forty people, Magnus's position devolved to that of a junior protocol officer, arranging the table seatings and assuring that the needs of guests from around the world were satisfied. He made note of the fact that German Jews were prominent at ministry dinners and receptions. "No one asked if they were Jews or not," he observed, adding somewhat to the contrary that "baptized Jews held official positions as civil servants."[38] Dazzled by the medieval ostentation of Wilhelm II's court, his native monarchism was now fortified by first-hand participation.

"The cultural level [*Kulturniveau*] of a nation is represented by its music and art," he insisted, and "indicative of the political level." In other words, a society that had produced Bach and Beethoven would naturally govern itself with commensurate dignity. Mass movements of any stripe were by definition inferior.[39] "Demagoguery and democracy are related words."[40] In Magnus's eyes, *kaiserliches Deutschland* (imperial Germany) before the Great War was unstained by corruption. Writers who created with "sharp and critical pens" a picture of abject militarism and *Junkertum* (Junker life) in Germany were distorting a history where subjugation to the king was honorable and as poorly understood by outsiders as the samurai cult in Japan. Seemingly blind to the fact that such abject monarchism was possible only in a rarefied stratum of society, Magnus ignored the widespread ridicule of Wilhelm's autocratic fits and bombastic oratory, especially after a bizarre interview with the kaiser published by the London *Daily Telegraph* in October 1908, in which he claimed to have instructed the British Army on how to end the Boer War.[41]

In 1909, at yet another dinner party, Magnus met a young woman "who looked like Rembrandt's Saskia, wearing a broad black velvet hat with white peacock feathers." They met seven more times, fittingly in the White Room of the kaiser's palace, and on the basis of this formal courtship were married in July 1910. She was Emmy Melitta Cecile von Quistorp, twenty-four years old, from Crenzow, whose family name traced back to the Swedish occupation of Pomerania in the seventeenth century. The von Quistorps had spawned Lutheran ministers and professors rather than soldiers. Nonetheless, Emmy's father, whose name was Wernher, had been a prominent landowner and her mother, Marie, was a von Below, an illustrious military name. She was thus firmly on a social par with the von Brauns.[42]

The skies ahead of them must have looked blue, indeed, with an open window to the most stratospheric levels of Berlin society. At minimum, it is safe to assume that Magnus was quite struck by his bride's physical presence, having had only seven chances to talk with her alone. At five feet and nine inches, she was nearly two inches taller and would have towered over him in a fashionable chapeau. She, too, had spent a year in London, at finishing school. And Emmy must have approved at least of her husband's promising career in the kaiser's government. Had they been situated anywhere else besides the bitter end of a gilded era, a misfortune which neither they nor anyone else could foresee, they might have reaped all the benefits of their station, which were considerable indeed.

Almost immediately, however, their tidy situation was altered when Delbruck moved from the trade ministry to serve elsewhere in a restless government and did not bother to extend his coattails to his assistant. In 1911, Magnus found no alternative but to leave Berlin altogether to be installed as *Landrat*, or district magistrate, based in Wirsitz, a town in the rural Prussian province of Posen—the German equivalent of Siberia.

Settling in a remote farming region where half the population was Polish, far from the glitter of the royal court that had so enchanted him and his bride, Magnus found solace in the independence of his new office but swallowed hard over his sudden exile to provincial life. To compound the change, Emmy gave birth to their first child, Sigismund, in April 1911. The venerable post of Landrat was traditionally a rock-solid sinecure that a young Junker might keep for the rest of his life. The reality, especially in hostile Slavic Posen, was less comforting. Magnus fretted that his initial salary of 3600 marks a year was no larger than his allowance as a student at Göttingen. The reaction of Emmy, whose stylish French hats would have to stay boxed in Wirsitz, was not recorded.

From November to January, the Landrat's duties consisted of going hunting with the local German gentry. With his Polish chauffeur toting his shotguns across the hay lands, Magnus sometimes killed ninety pheasants and sixty hares in a single day, much in the rapacious spirit of the Teutonic Knights. As the official in charge of taxation, such comradery was useful in seeing behind closed doors, much as a round of golf might be for future generations. He also waved his pen at the construction of local railways and roads, oversaw the police, and made plans for wartime mobilization.

The relationship of a German administrator to Polish landowners was, of course, problematic.[43] There was no real social intercourse, even through the mutually revered sport of hunting. The farmers and townspeople had created a network of their own social and economic organizations that amounted to a parallel state. Since the founding of the German Reich in 1871, the Prussian government had tried to Germanize the Polish zone by making German the language of instruction in the schools and by settling Germans on the land, but the administrative networks of the Poles enabled them to sidestep these intrusions. The great mass of the Polish middle classes, farmers, and workers were loyal to the state and gladly made use of the institutions of the Prussian Army and civilian government, but inwardly they rejected it.[44]

When Magnus discovered in his predecessor's files a long list of Poles who were to be arrested in case of war, he wisely protested to superiors that its existence was inflammatory. But when he learned that a local Polish duchess was opening her library to young nationalists, he forced her to stop. Though since 1908 land owned by Poles could be legally sold to Germans by fiat, under a kind of eminent domain, German Landrats in the region did not feel entirely safe "on their own soil," as von Braun put it. At best, he was respected grudgingly by the vassal Poles as an overlord, a representative of an occupying power. This tension no doubt added to the misery of a post far from the ballrooms of Berlin and Potsdam.

On March 23, 1912, Emmy gave birth to their second son, whom they called Wernher after her father.[45] The next two years, as Europe spun toward war, were relatively stable and uneventful for the family, their distance from posturing capitals perhaps something of a blessing. All in all, it was a peaceful interlude to be remembered fondly in later life after time and truly ugly circumstances had sanded away the rough edges. Much in the mold of generations past, Magnus became a remote father who commanded respect, a man whose official persona claimed all of his attention. The children were his wife's concern. His second son, who strongly resembled her, would in particular always be a mystery to him, or so he claimed in old age, if only because Wernher would grow more and more enamored of speculation than was ever condoned by Junker code.

2

MEMORIES OF DEFEAT

TIME TENDS TO erase the prosaic elements of a great war's beginning, replacing them with scholarly concerns. Who will remember the ordinary lives that were ruined without ever firing a shot? Yet for most of those in the path of clashing armies, grandfather's whereabouts are a more urgent issue than the strategy of renowned generals. The von Braun's native soil lay across one of Europe's hinterlands where national borders never seemed to coalesce—a corridor of mutually distrustful peoples, a favorite invasion route between rival powers.[1] Germans, Poles, Latvians, Russians—history of course belonged to whoever made it.

There is still today a palpable feeling of primitiveness about these lands that end at the Baltic Sea. The weather is grim most of the year. Reconciling German and Slavic pride has never gone well. Like stones stuck in the earth that can be driven over or around but not pried out, body mass has ensured survival. Living here could be like standing on earth that might turn to sand at any moment, swallowing up possessions that may or may not get regurgitated later under a different set of laws, an alien language, a new overlord. Again and again, people have learned how to pick up the pieces and start anew.

Magnus and Emmy von Braun heard from their chauffeur about the death of Archduke Francis Ferdinand, Austria-Hungary's portly heir apparent, one day after a Serbian student named Gavrilo Princip shot him and his wife in the Bosnian capital of Sarajevo on June 28, 1914. A month later, as the mustering of more than a million Russian troops accelerated the domino collision of alliances tumbling toward war, the young Landrat proudly assumed his military role of German mobilization officer. It was time to put aside the ceremonial trappings and accomplish the real task. For Germans of nearly every class and political persuasion, enthusiasm for war—a defensive action above all, they believed, against Czarist Russia—translated into a sense of glorious escape from the peevishness, the *Verdrossenheit*, of ordinary life.[2] That most people under his authority did not perceive themselves as German in any way would have to be overlooked.[3]

Under existing war plans, which called for the quick annihilation of France, Germany would not invade Russia.[4] The remote front covering East Prussia was sparsely held by only one cavalry and eleven infantry divisions totaling about 135,000 men, who were woefully deficient in heavy artillery, medical service, and field telephones. But what would such weaknesses matter when victory was assured in a few months? Already massing against them faster than Russians had ever moved before were eight cavalry and thirty infantry divisions consisting of some 650,000 men, albeit even more ill-prepared.[5] Great expanses of human flesh thus faced off with little more than ignorance and patriotism to propel them forward—perfect conditions for an abattoir.

Under these circumstances, the Landrat's duties were superfluous, though "the holy moment" and his Prussian sense of discipline would have compelled him to orchestrate the winding of cuckoo clocks if it had been part of written orders.[6] Wirsitz, which lay west and south of the Vistula River—well outside the natural path for invaders across East Prussia—was on relatively secure ground, though any proximity to Poland was discomfiting. For now, western Poland was quiet, with the Russian offensive confined to frontier raids. Up in Neucken, however, 150 miles to the northeast, Magnus's elderly parents wisely abandoned the family estate and fled in a carriage to his sister's house on the Baltic shore. They had evacuated twice before, in 1807 and 1812, but it was no less traumatic this time.

For Emmy, it was an especially terrifying moment. She knew that her husband was ever more vulnerable as a minion of the German government. Their upbringing would reflexively stiffen their spines—part of the reason they and others of their class won such postings in the first place. But for the same reason they would also stand out as targets. As for the boys, while they might not comprehend their situation, they were still at an age of raw vulnerability to a sense of peril at home. It would stay with them forever, even if they could not yet put words on it.

In retrospect, the significance of the battles that ensued went well beyond military equations. Compared to the flesh pits to come, they were mere skirmishes. Nonetheless, the machine gun and artillery slaughter spawned heroes, though they were tarnished by the carnage itself. Generals and statesmen learned quickly how to rationalize a scale of acceptable casualties that would have seemed apocalyptic only a few years earlier. General Max von Prittwitz und Gaffron, called the "Fat Soldier" by his troops, a nervous incompetent with no blood ties to the soil of East Prussia, commanded the German forces. When the Russian army crossed the Prussian-Polish frontier—a wilderness of forests, bogs, and lakes—in mid-August, von Prittwitz decided to wait for the terrain to slow them down and funnel them into a location more central to the planned battle.[7] But this meant giving up the extreme eastern corner of East Prussia without a fight, an unthinkable scenario to one of his corps commanders, General Hermann von François, a native Prussian. Von François surged his troops forward to challenge the first Russians over the border, spoiling von Prittwitz's textbook strategy.

Von François stalled the invaders for a few days, with both sides suffering ghastly losses. A reluctant von Prittwitz then decided that time was running out and threw his divisions into an attack at Gumbinnen, about 75 miles east of Neucken, on August 20. The resultant battle was a costly draw, like so many horrors to follow. One German corps lost 8000 men in several hours. The Russians suffered some 17,000 casualties. A flustered von Prittwitz panicked when he learned that more Russians were advancing from the south. Hysterical, shut up alone in his office while madly telephoning superiors in Coblenz to beg their permission to retreat behind the Vistula, he was promptly sacked. As his replacement, the General Staff called out of retirement a stolid sixty-seven-year-old Junker general named Paul Ludwig Hans Anton von Hindenburg und Beneckendorff.[8]

In retrospect, it was the kind of off-beat selection that would make later generations wonder at the way Fate trifles with lives. An obscure has-been thus began rising to take a pivotal role in his nation's eventual destruction. By the end of August, Hindenburg—in reality little more than a figure-head—had extracted a kind of victory from the blood, elating his coun-trymen and inflating his worth as a political figure. A quintessential product of the Prussian ruling class who had fought in the Austro-Prussian war of 1866, he would soon come to incarnate all that was deemed glori-ous in the German people, towering like Bismarck over the fatherland as a superhuman national hero.[9] The first summer of war had already brought failures on the Marne and Aisne rivers along the western front in France—appalling battles involving millions of men and thousands of guns in concentrations never seen before. Hindenburg's slight success was duly exaggerated. Never mind that the ten-mile-wide trap he managed to lay for the Russian regiments turned into a slough of dead men and horses, that German soldiers went mad as the gore deepened, firing their artillery into seething masses of Russians without even taking sight. When it was over, German newspapers trumpeted the propaganda spoon-fed to them by the Reichswehr, reminding readers that 500 years earlier, in 1410, the Poles had virtually wiped out the Teutonic Knights on the same battle-field. Now, at last, those progenitors had been avenged.

When Magnus's eighty-one-year-old father returned to the estate at Neucken in mid-September, he found that the Russians, whose scout units spent only a day there, had let the place be. The "inevitable lobs," as Mag-nus put it, caused some minor damage, but most of the family treasures had been safely hidden by a faithful housekeeper. Forty years later the Russians would return with a vengeance, but for now the von Brauns' beloved Heimat was intact.

In November, Wirsitz was threatened directly by Russian forces rolling toward Posen and Silesia, Prussia's industrial heart, after pushing Hindenburg's troops away from Warsaw.[10] In the midst of the crisis, Magnus sent Emmy with 3 million marks from the local bank to the Deutsche Bank in Berlin—a two-day journey of considerable risk.[11] Believing it would cause panic if she were known to flee the region with Wernher and Sigismund, he kept his family in Wirsitz rather than send them back to the capital. Had he or any other civilian been fully informed about the level of carnage wrought by modern warfare, it seems unlikely he would have made such a quaint attempt to maintain morale. Magnus turned his attention to drilling young reservists while Emmy worked at the other end of the meat grinder, helping to organize a military hospital, which quickly overflowed with wounded soldiers. Their sons, aged two and three, were cared for by servants, presumably Polish.

Hindenburg, meanwhile, had been elevated to "Commander in Chief, East" by the kaiser, and his legend continued to mushroom. As winter weather enforced a general stalemate, with some German units losing a third of their strength to frostbite, Hindenburg began to understand how one could win battles but fail strategically. He was hardly alone in this old paradox, of course, but given his later role in German politics it seems worth noting his experience here. It may also be worth pondering that this Prussian hero did not flinch from introducing one of the war's most heinous new weapons. In January 1915, 18,000 special artillery shells developed by German Army researchers under the direction of the illustrious chemist Fritz Haber were delivered to Hindenburg's regiments.[12] They were added to the barrage against the Russian invaders, but the poisonous gas they contained proved useless in subzero temperatures. Nonetheless, Germans and Slavs alike continued to die like dogs—after only five months of fighting, there were more than one and a half million dead, wounded, or captured on both sides—while it slowly dawned on people at home that this war would be different.

IN THE SPRING of 1915, Berlin quietly sent orders to its Landrats in agricultural regions to estimate how much grain was held in reserves. As the squeeze on the German economy tightened, people began to comprehend all too well the meaning of total war.[13] Ration cards had been issued in

January, limiting each individual to 4.5 pounds of rough *Kriegsbrot*, or war bread, per week. Chocolate, coffee, and tea were already disappearing from store shelves. Emmy von Braun, showing the kind of commonsense insight her husband often seemed blinkered against, warned him that the government's query was double-edged, in that it could bring reprisals against the local population if anyone provided inaccurate or false estimates suggestive of black-market activity.[14] Her opinions must have carried considerable weight in the household, because Magnus immediately boarded a train for Berlin to report this concern as his own to superiors at the Ministry of Interior.

A week later, Magnus received a telegram from a cousin in the Ministry, telling him that Clemens von Delbruck, who as secretary of the Interior had risen to head the most important civil agency in the Reich, wanted to bring him back as his aide. It seems likely that this was Emmy's desire all along, hence her husband's personal trip to Berlin on a matter that could have been handled through channels. Magnus no doubt saw his Landrat position in Posen as an inviolable duty to the kaiser, a charge to be carried out with Prussian rectitude. That it placed his family in harm's way was part of the sacrifice any Junker would bravely accept. He later claimed that the decision to leave Wirsitz was very difficult for him. Emmy, on the other hand, saw two children frightened by the proximity of war in a region where ethnic hostilities festered just below the surface even in peacetime. Chancellor Theobald Bethmann-Hollweg had already accepted recommendations to protect Prussia by creating a "frontier strip" in northern Poland from which Poles and Jews would be resettled eastward and replaced with Germans—the first time in modern European history such forced migration was considered an acceptable solution for national and ethnic conflicts.[15] Though it was never implemented, even its planning might have entangled the family in atrocities.

Emmy also was perceptive enough to realize that Magnus's duties were in reality of little consequence. Whether she confronted him on these terms or steered the situation more subtly cannot be known, but the outcome—a return to their beloved Berlin, surely the most reassuring place to be as the war roiled on—demonstrated her strong hand.

Magnus resumed the workaday functions of a personal secretary for Delbruck that he had lost five years earlier. Interior was a labyrinthine and cumbersome organization, with Delbruck also acting as Bethmann-Hollweg's deputy. His primary concern at the time was where to draw the line between free markets and "war socialism," an economy directed by the state for military production. None of the major powers had ever faced this question so starkly. No formal plans had been made for the management of Germany's wartime economy, because the prevalent delusion before 1914 was that modern wars would be brief. In the past, war had been limited enough to be waged without reordering national priorities. But it was increasingly clear that warfare in an age of industrial technology required total commitment of all the nation's resources, human or otherwise.

In this regard, Germany began the war gravely handicapped. Its food stocks and raw materials were more limited than the Allies', compelling German officials to extend their war organization deeper into society.[16] Food was already becoming scarce, a misery once confined to peasants caught in the crossfire between rival princes.[17] For the first time, the government took possession of potato and grain crops as soon as they were harvested, paying farmers a fee to cover expenses.[18] Food shortages were leading to hunger riots, black marketeering, wildcat strikes, and plundering of shops.[19] In addition, virulent inflation further pitted rich against poor, town against country.[20]

As in 1910, Magnus's social skills, at least within the bureaucracy, became of most value to his unpopular boss. As the government took over rationing, distribution, and price control first of bread and then an ever-increasing number of basic foods, Delbruck was hounded by the press over the war's domestic cost and felt the acute need to handle his public statements as gingerly as the army treated accurate information from the trenches. Ad hoc agencies did not work well and prices kept rising. Producers hid large parts of their output and sold it on the black market for obscene prices. Von Braun became what in a later age would be called a press secretary, answering questions from "the monster," as he called Berlin newspaper reporters, with all the aristocratic aloofness he had been taught since birth. It was necessary, he felt, to give the public some plausible motives for the often clumsy and repressive measures being taken to control the economy.

To this end, he organized a news room at the ministry, employing "educated girls" to read and clip the daily papers for distribution around the government. The operation also wrote its own articles, which were planted in foreign newspapers as "reports from Berne."[21] In other words, it was a classic propaganda machine and counter-intelligence effort aimed both inside and outside Germany. The ability to manage the flow of information, thereby indirectly affecting the formation of policy by those above him, fascinated von Braun and became a central element of his political consciousness. He eagerly continued at this post after Delbruck was forced to retire because of poor health in June 1916. This was the year of macabre battles at Verdun and the Somme, so his talents as a spin doctor were sorely needed.[22]

Delbruck was replaced by Karl Helfferich, an economist who had been director of the Deutsche Bank.[23] Helfferich soon clashed with the military over war production, especially after Verdun and the Somme created weapons shortages.[24] As the balance of power between military and civilian leadership became a more and more volatile issue, von Braun tried to push Helfferich into responding to army critics through the press. But Helfferich did not want to speak on unorthodox channels—a political prudence that von Braun interpreted as weakness. As for Bethmann-Hollweg, von Braun felt that the chancellor did not live up to the "strong masculine nature" supposedly advocated by Nietzsche.[25] Helfferich's willpower was being sapped by "that old bloodsucker, that spider, Skepticism." All in all, von Braun began to despise what he felt were the "soft politics" of his superiors.

In December 1916, General Erich Ludendorff, who was now running the entire German war effort with Hindenburg like two Japanese shoguns, made it known that he wished to set up a centralized press agency representing the government, especially since they were having trouble enforcing their strident military censorship. The German public had been kept largely uninformed about the war's disasters and still expected victory.[26] The civilian leadership, however, was against surrendering yet more power to the generals, who were running a smear campaign against them (even the kaiser had resisted appointing the Hindenburg-Ludendorff duumvirate, because it would take most of his remaining power). As malnutrition reached crisis levels during the "turnip winter" of 1916–17, while military casualties passed the 4 million mark with no end in sight, the government was under increasing pressure for electoral reform. Basically, the wartime rulers had broken faith with their subjects.

Helfferich himself came to favor the notion of democratic voting rights. The kaiser's pathetic behavior in the face of disaster—champagne dinners, impatience with real-life military discomfiture—helped fuel the public's radicalization. But von Braun opposed any movement toward democracy, making his objections to reform known to Helfferich and throughout the Ministry. Almost as out of touch as the kaiser, he wrote to the Austrian foreign minister in April 1917 that "the recent unrest in Germany, especially the strike in the munitions factories, must be traced to the Russian revolution. The unrest is believed to have been caused not so much by famine and need as by political motives and the spirit of internationalism."[27]

He also joined the opposition to Bethmann-Hollweg's continuing as chancellor. By the summer of 1917, with the nation in a state of total war, Bethmann was a broken man, his diplomatic and domestic policies in a shambles, his attempts at compromise all turned against him. His fatal mistake was to approve a peace resolution put forth by Social Democrats and Centrists in the Reichstag, triggering Hindenburg and Ludendorff to threaten the kaiser with their resignation. Bethmann knew he could not withstand them and resigned first, though the kaiser was not pleased with the implications for his own power: his top two generals now held complete sway in political affairs.

Von Braun took it upon himself ("an expensive game" he called it) to toss into press, Reichstag, and military circles the name of Georg Michaelis, an obscure Prussian civil servant.[28] With hearty support from Hindenburg and Ludendorff, and a paucity of other feasible candidates, Michaelis—sixty years old, deaf, pietistic, a total novice in high politics—

became chancellor of the Reich. That he knew nothing of foreign affairs and had never dealt with the Reichstag was apparently irrelevant; he appealed to the soldiers, whose lower ranks were rife with disillusionment.[29] And he was a flyspeck to Ludendorff.

By the autumn of 1917, as the monarchy itself and all of its bureaucratic trappings began to splinter under the weight of an unwinnable war, von Braun's favor was repaid when a press office was established for the first time under Michaelis's aegis.[30] Ludendorff still ran his own military press operation in a desperate attempt to raise morale, but as tension increased between Germany's military and civilian leaders, the two were often at cross-purposes. He kept up his pressure for a unified office, but the chancellor maintained his dead-set opposition.

Von Braun's new title clearly indicated whom he was expected to serve: "Emperor's Director in the Reich Chancellory and Chancellor's Press Chief" (*Kaiserlicher Direktor in der Reichkanzlei und Pressechef des Reichskanzlers*). Here, finally, was a position of real power, not a Junker sinecure. He held daily meetings with the chancellor himself, a heady experience, though Michaelis's ineptitude was already widely apparent. He felt that the press as a whole welcomed his role, but also knew that he had to play favorites among the country's major editors to properly sow government policy. For example, the editor-in-chief of the liberal, Jewish-owned *Berliner Tageblatt*, Theodor Wolff, had a "horrible hate of the government," while Georg Bernhard of the conservative *Vossischen Zeitung* "always had political understanding" and thus "could always be relied upon for cooperation and discretion." As the political calculus in Berlin became more and more abstruse, however, these facile categories worked against him.[31]

Von Braun was no match for Ludendorff and Hindenburg, of course, who during the course of the war dragooned into submission almost every category of German authority, from kaiser to trade unionist. The military, hell-bent on seeing the war through to victory at any cost, bridled whenever nonmilitary leaders spoke via the press about the postwar future of Belgium, say, or Poland. Von Braun, like almost everyone else at this juncture, found Kaiser Wilhelm to be out of touch, yet also chaffed against Ludendorff's insistence on force to solve matters that might better be addressed through diplomacy.[32] Ludendorff was thus sharply aware that von Braun was standing in his road. "Undersecretary von Braun should get out of our way," he wrote bluntly to the chancellor, who still saw no profit in throwing von Braun to the lions.[33]

As popular pressure for peace burgeoned and the military leadership drew farther and farther away from the Reich chancellory, von Braun's position became steadily more precarious. Whenever Theodor Wolff, who was distrusted by both factions, published an article critical of the military, Ludendorff thundered that von Braun was to blame. In a few cases, acting too independently, he no doubt was. Von Braun's superiors ineluctably lost their nerve in a treacherous battle of press releases. Debate centered on whether the government should openly renounce an offending news article as not representing its position or von Braun should take personal responsibility. The frustrated press soon began to question whether von Braun's office was doing any good at all.

Inevitably, von Braun's enemies outnumbered his supporters. When Michaelis stepped down from office to everyone's relief at the end of October 1917, von Braun lost his only patron. He resigned during the first week of November and was ordered on New Year's Day, 1918, to a military administrative job on an island in the Gulf of Riga, between Lithuania and Latvia. This was utter banishment, indicative of the true cost of the game he had played badly and lost. He had no choice but to leave Emmy and the children with her relatives on the family estate at Crenzow, near the Baltic Sea coast, where they had been living for some months to escape the worsening conditions in Berlin. A Russian-style horse-drawn sled driven by a man wrapped in furs took him the last 70 miles to a ferry in Lithuania.

"You couldn't think of a place where you felt more exiled," he wrote of the desolate outpost, displacing Wirsitz in that category. For three months of winter, it was cut off from the mainland by ice. Public life was thankless, he lamented, believing like so many others that he had only been doing his duty. Six weeks later he moved to a German fort in occupied Russia, hardly an improvement. One of his first tasks was to dispose of three thousand starving horses. They were herded into an open ditch, where crows picked their bones clean by springtime.

The political news from Berlin grew bleaker as the magnitude of impending defeat became apparent well beyond military circles, despite Ludendorff's censorship. "This period belongs to the most horrible memories of my life," von Braun wrote. He compared the inevitable collapse of the German monarchy to a centuries-old house burning to the ground in the middle of an earthquake. Constitutional reforms were no longer sufficient to satisfy the masses, who demanded an end to their misery. The urban population was starving or dying of influenza and tuberculosis. The farmers were outraged by the meddling of government agencies in their work.[34] After the failure of 1918's spring offensive, the army began literally to fall apart from desertion and insubordination. Even the terror of a giant new artillery piece firing on Paris in March from a distance of 75 miles could not stem the tide of defeat. New American divisions on the western front left the final outcome in little doubt—the prospect of a fifth winter of war and social chaos was untenable. At the end of September, Ludendorff finally admitted to himself what was obvious to everyone and called for an armistice "to avoid a catastrophe," which was already at hand.[35]

Exhausted and nerve-wracked, Ludendorff resigned a month later. A new constitution limited the kaiser's monarchical power to a level comparable to the English crown. But reforms were overtaken when a leftist revolution, sparked specifically by naval mutiny in the port city of Kiel and generally by a desperate desire for peace, brought on Wilhelm's abdication on November 9. He fled to Holland, triggering the demise of German dynasties everywhere. Two days later, Reichstag leaders signed an armistice agreement at Compiègne.[36] Friedrich Ebert, the new Social Democratic chancellor, announced a "Government of the people" that would "make every effort to secure in the quickest possible time peace for the German people and consolidate the liberty which they have won."[37] Prevention of civil war and famine were his top concerns. When the first front-line troops marched back to Berlin in December, many in the masses of onlookers wept at the decrepit spectacle. Along Unter den Linden, the professional *Flittchens* (floozies) soon had to compete with war-widowed housewives.

Although the end of fighting was a blessing, Germany found itself on the brink of complete internal collapse. Starvation remained widespread and coal was scarce, not only leaving homes without heat but also hobbling industrial production just when it was most needed. Millions of exhausted soldiers faced sudden and largely uncontrolled demobilization.[38] The government had financed the war almost entirely by issuing debt, which now amounted to half the national wealth.[39] On the first of January 1919, as civil conflict flared around the country, the Prussian Ministry of Interior sent von Braun to Stettin (now Szczecin), a shipbuilding center near the Baltic coast, to stand in for a sick police chief. On a perilous assignment under anarchic circumstances, he reported that Communist workers had placed machine guns around the police headquarters. They failed to attack, probably because of opposition by radical right-wing *Freikorps* paramilitary units that crushed similar revolts in cities across northern Germany. Assuming that this confrontation between a middle-aged Junker monarchist and Spartacist youths actually took place— old-line civil servants were doing their best to compile stories demonstrating the damage done by leftists—it was surely the high point of a bizarre interlude in Stettin.[40]

Von Braun's old-boy network soon enabled him to "fall upstairs," as he frankly put it, to become chief administrator in the East Prussian town of Gumbinnen, where the Teutonic Knights had been avenged by Hindenburg back in the glorious summer of 1914. Though the Prussian government was now nominally socialist, it was pragmatic enough to recognize the value of some continuity at the local level. The brutal suppression of worker uprisings by *Freikorps* in Berlin finally opened the way for elections to form a national assembly, which convened in the first week of February 1919. Because the capital still seemed too tumultuous to guarantee an undisturbed meeting of the new parliament, the provincial city of Weimar was chosen instead, thus lending its name to a republic that would eventually spell disaster for the elder von Braun even as it fostered his second son's success.

The leftist revolts throughout Germany against the tattered remnants of Wilhelmine order would provide a generation of conservatives with a potent myth of Bolshevist, especially Jewish Bolshevist, conspiracy. That the uprisings were spawned mainly by common misery and war-weariness would be lost in anti-Communist propaganda during the Weimar era and beyond. A savage battle, verging on civil war, over the establishment of a short-lived Soviet-style government in Munich provided especially fertile ground. Some of the revolutionary leaders happened to be Jewish, and this was enough to stoke an atmosphere of irrational fear already spreading from the Russian Revolution to every level of German society. Bavaria would soon become a nest for the radical right, including not only paramilitary groups but also the Reichswehr itself.[41]

Indeed, much of the military, judiciary, police, and government bureaucracy would remain hostile to the Weimar Republic, which was denounced by many right-wingers as a *Judenrepublik*.[42] For the von Braun family, life under democratic socialism was nonetheless infinitely more comfortable than the last days of their beloved monarchy. They were reunited in a thirty-six-room mansion where tame elk roamed the lawns. Magnus's main job, as he saw it, was to guard against the importation of Bolshevism (widely demonized as the embodiment of Jewish socialism) from Lithuania and Latvia as German soldiers, many of them deserters, straggled home from the Eastern front.[43]

Since he had no practical way to accomplish this task, his own existence as a representative of true German virtue became his best and only weapon. In 1919, the bleakest year in modern German history next to 1945, he and Emmy expanded their family with a third son, whom they called Magnus. Seven-year-old Wernher, a beautiful boy in white sailor suit, enrolled in first grade at the Friedrichsschule in Gumbinnen. Thus, in the midst of chaos, sprouted an outpost of *Junkertum*.

As the new decade opened cautiously, the new republic was rocked by a right-wing coup attempt in March 1920. Mutinous army brigades and Freikorps, with many soldiers displaying a black *Hakenkreuz* on their helmets, marched through the Brandenburg Tor and occupied the government quarter of Berlin. The political wirepuller was Wolfgang Kapp, demagogic founder of the ultra-nationalist Fatherland Party during the war, who now declared himself chancellor.[44] To all those hostile to the Republican developments at Weimar, the mere arrival of Kapp's reactionary forces raised hopes for the return of the hallowed past.

Freiherr von Braun was among them. Only a few days before the putsch, he had seen Kapp, whom he had once known as an agricultural official (*Generallandschaftsdirektor*) in East Prussia, at a restaurant in Berlin. He learned of the putsch by telegram from East Prussia's ranking civil service head and, believing that Weimar was crumbling, took the sympathetic step of declaring an autonomous "Republic of Gumbinnen" from his 36-room mansion. He never had a chance to define just what his microstate would be, however. When the commanding Reichswehr general in the region, who like most of the military elite regarded Kapp as a parvenu, heard of this creative action, he quickly sacked von Braun and thirteen local magistrates, adding a stiff fine for good measure.[45] Kapp's ill-conceived adventure ended in a week after a general workers' strike halted all public services and he proved unable to crack the high ministerial bureaucracy. He subsequently fled to Sweden.

There ended von Braun's first incarnation in government. Another dozen tortuous years would pass before he was called upon again to play a far more dangerous role. Always able to tap his social milieu, he was soon appointed by Hermann Dietrich, a banker and Reichstag vice president (and uncle of actress Marlene Dietrich), to be director of the Raiffeisen agricultural cooperative bank, or credit union, for Brandenburg and Schleswig-Holstein.[46]

The mental state of his family when they moved back to the capital in the summer of 1920 is not hard to imagine. They had survived a national catastrophe in some of its worst venues. Fortunately for them, perhaps, the entire country was under far worse duress, with tribulations ahead that would only increase widespread nostalgia for Junker rectitude. They thus began slowly to rise again to the top, where *Pater* von Braun knew they belonged.

Left to right: *Sigismund, Werner, and Magnus von Braun, Berlin, 1924.*

3

"HIGHLY TECHNOLOGICAL ROMANTICISM"

ISTORIANS MAY NEVER possess an entirely rational, cohesive explanation for why the Great War happened. There was none at the
time. The carnage was so much deeper than any set of issues put forth
as justification that the generations who survived seemed either "lost" or
bent on revenge.[1] About 2 million German men were killed and 4.2 million wounded, amounting to nearly one-fifth of the male population.
Many commentators have noted that the political myths that soon
shrouded the war resonated most strongly among those who were too
young to have actually fought in it.[2] For this large cohort born in the
first decade or so of the twentieth century, it was therefore plausible to
glorify war as unambiguously positive, despite the hardships they might
have known at home. By the early 1930s, as praise grew for the heroism
of the "front generation," the younger "war-youth generation" responded
out of guilt or admiration to supply many of the Nazi movement's
activists.

What is clear is that the war's "orgy of senseless violence" drew a bloody
line between the mannered past and the technological future.[3] Technology perhaps appeared to some to offer a relatively pristine road away from
disillusionment—a new kind of certainty to replace the obliterated cer-
tainties of the pre-1914 world—though this path was as prone to corrup-

tion as any other. For Germany, in any case, the future could hardly have turned out worse. A mere fifteen years lay between the debacle of 1918 and Hitler's ascension to the chancellorship in 1933.

From the war's desolate aftermath rose an interest in one particular technology that would grow at the margins of society, much as aviation had done before the war, and then also flower hugely under military sponsorship. Like aeronautics, rocketry in its early days possessed a romantic aura that attracted more than its share of visionaries and fools. For a long time its promise was far more imaginary than real, appealing especially to young men who did not fit comfortably into established enterprises. In postwar Germany, little was left of anything that could once have been called established, leaving more room than usual for eccentrics—or "humbugs, charlatans, and scientific cranks" as one contemporary authority observed—to be taken seriously.[4]

At this juncture, rockets were of marginal interest as weapons, having been supplanted by accurate artillery a generation earlier. Without certain provisions of the Treaty of Versailles, the military might have stayed away from the whole subject for another generation.[5] Often called a peace so vengeful that vengeance was its only assured outcome, the treaty placed sole guilt for the war on Germany, ceded away a tenth of its population and 13 percent of its territory (mostly in the Alsace and Lorraine regions on the border with France, and Prussia on the Polish border), saddled the fragile new republic with unspecified reparation payments that could reach more than 130 billion gold marks, plundered the nation's colonial empire, and cut its military to an internal security force with token navy, no aviation, and a ban on conscription.

This was the *Knechtschaft*, or slavery, that the Nazi's "Horst Wessel" anthem swore to end, but almost all Germans were passionately against the treaty. Disarmament was hardly unjustified, yet voices across the political spectrum reacted with outrage and disbelief. In spite of all its horrors, the war did not exhaust nationalist fervor. Ubiquitous opposition to the treaty served to stifle criticism of Wilhemine attitudes and fuel suspicion of parliamentary democracy.[6] The inevitable response of unrepentant German militarists was to exploit whatever loopholes were available when they were not simply ignoring the Versailles *diktat* through overt or covert weapons development.[7]

Among the treaty's detailed tables stipulating how many munitions the postwar Reichswehr could possess (52,000 revolvers, 30,000 sabers, etc.), rockets were not even listed. Exactly eighty-four 10.5-cm howitzers were allowed, a derisive figure to field officers who had just surrendered nearly 60,000 heavy-gun barrels to the Inter-Allied Commissions of Control.[8] Since these leftovers were obviously unsuitable for a new army, the military leadership gradually concentrated on building up a *Zukunftsarmee*, or army of the future.[9] The army of the past, after all, had completely collapsed at the end of 1918. They would soon realize the potential for rockets to deliver mayhem now unavailable to them through traditional means.

Socially, the treaty turned the wartime Reichswehr into a professional army more aloof than ever from civilians, with a monarchist and ultra-conservative officer corps still drawn from the old imperial military caste and the rank-and-file from hard-bitten anti-Republican Freikorps elements. It was a dangerous combination of royalty and rabble, united by venomous hatred of Bolsheviks, Social Democrats, and Weimar utopians (often meaning Jews, specifically). Though the high command discouraged political activism out of concern for strict discipline, in practice there was seamless adherence to rightist dogma. Collaboration with patriotic paramilitary gangs such as the *Stahlhelm* (Steel Helmet), *Oberland*, and *Werewolf* was taken for granted as a way to skirt the treaty's limitations.[10] Throughout the 1920s, there could be no mistaking the Reichswehr's sympathy for radical nationalists or its hostility toward the Weimar government.

The treaty also turned broad segments of the German middle class against democracy. President Woodrow Wilson had enjoyed a brief period of respect after the German government presented his Fourteen Points to the public as a reasonable basis for settlement.[11] When the punitive reality of Versailles settled in, however, so did revulsion against Wilsonism. With the upper classes still strongly monarchist and the proletariat enamored either of Lenin on the left or Freikorps heroes on the right, the stage was being set for a leader who could restore order at any cost. In June 1920, the three mainstay political parties of the new Weimar system—the Marxist Social Democrats (oriented toward industrial workers), the liberal German Democrats (middle-class professionals), and the conservative Catholic Center Party—failed to win even a simple majority in national elections for seats in the republic's first Reichstag. The coalition would remain in the minority, outnumbered by hostile deputies on the right and left.

For the von Braun family, like millions of other Germans on every rung of the social ladder, it was a time of looking for stability where there was none. Because of his tangential association with the Kapp putsch, father Magnus was essentially disqualified from political life at the national level even if his hardcore monarchism were overlooked. He devoted himself to the relatively mundane work of meeting in "Christian solidarity" with farmers and arranging loans through the Raiffeisen credit union.[12] For a man of his station this was humble toil, but it was a steady, respectable calling when few were to be had anywhere.[13] It provided Emmy and her three boys with a semblance of normality, which was of inestimable value under the circumstances. In 1920, the family moved from Gumbinnen to 20 Tiergartenstrasse in Berlin.

Two other societies whose citizenry suffered the Great War to varying degrees and were trying to move forward would form separate seedbeds for rocketry: Russia and the United States. Germany and Russia were wracked by revolutions that nonetheless helped create fertile conditions for new technology. America was the least disrupted and would take up the subject, beyond the private efforts of one lone experimenter, Robert Goddard, only after another worldwide conflict. These three national strands, though separate at first, were fated to intertwine—sometimes loosely, sometimes tightly enough to form a single cord. The middle von Braun son, Wernher, would become entangled in all three through the improbable course of his life. At this moment, however, he was still an aristocratic schoolboy of no special promise, attending Berlin's elite French Gymnasium school founded by Frederick the Great, in the exclusive Charlottenburg district.

Among these sociopolitical threads was an intellectual element that formed a backdrop for the early development of German rocketry, albeit at considerable distance. Beginning around the turn of the century, professors at German engineering colleges (such as the Technical University of Berlin in Charlottenburg) developed a novel way of thinking about machines, material progress, and culture—*Technik und Kultur*. Novelist Thomas Mann later called it a "highly technological Romanticism."[14] What American historian Jeffrey Herf has more recently labeled "reactionary modernism" led to a postwar right-wing cult of technics and, ultimately, the nazification of German engineering. Along the way, during the Weimar decade, it helped open a window for a marginal, nascent technology to be nurtured by the state.

The central tenet of this ideology was that Germany could be both technologically advanced and true to its old folkloric ideals. Accordingly, the romanticized *Volk* needed to be protected from the corruption of Western *Zivilisation*, especially the degenerate *Amerikanismus* whereby financial profits determined the path of progress. To accomplish this, rationality would have to mix with mythology, contrary to the Enlightenment notion that the two are mutually exclusive. The macho rhetoric of the day, primed by bitter memories of the war, called it "thinking with the blood."

In hundreds of books, lectures, and essays from the technical universities and nontechnical proto-Nazi writers such as Ernst Jünger—whose novel about the war, *In Stahlgewittern* (Storm of Steel), which glorified front-line military life, was much admired by Hitler—and Oswald Spengler—whose most famous work, *Der Untergang des Abendlandes* (The Decline of the West), claimed that "war is the creator of all great things"—and from all across the political spectrum, *die Streit um die Technik* (the debate about technology) examined the relationship between Germany's soul and modern technology. Rather than a backward-looking Luddism, it fostered a widespread belief that technology was indispensable to German nationalism. Though a minority of the 300,000 engineers in Weimar Germany probably concerned themselves with this debate, it permeated the profession's elite and defined the terms of the day for anyone who contemplated the problems of technology, society, and culture.[15]

Against this backdrop, interest in rocketry arose at a very low level. Hindsight bias no doubt aggrandizes the lives of a few itinerant theorists and experimenters—even using these labels serves to enlarge otherwise obscure individuals. The aviation pioneers of America and Europe, to whom they are sometimes compared (and to whom Wernher von Braun compared himself), [16] produced far more authentic advances. It is unlikely that any of the characters now noted for seminal activities in rocketry during the 1920s would have been remembered if not for the enormous wave of military interest that came later and left nearly all except von Braun in the mist. Certainly there was no other organization of state or commerce besides the military capable of transforming science-fictional dreams into reality of any scale beyond the backyard, as is still largely the case today.[17] Though it is conceivable that the flying-machine inventors of the early twentieth century might have developed an air transportation industry on their own without military investment—though this is not, of course, what happened—there was no chance of such commercial independence for the rocketeers.

What was eventually accomplished under military auspices alone makes it tempting to mount a whiggist search for forerunner personalities.[18] This exercise in retrospective rationalization has been especially appealing to American journalists and historians who have cosmeticized the reputations of German rocketry pioneers, especially von Braun, by casting them in whole or in part as apolitical space-travel enthusiasts, rather than ideologically branded weapon-builders. Anti-Communist hysteria in the United States after World War II helped to whitewash the past of men who could have been prosecuted under the Nuremburg Code if they had not been enlisted in the technological race against the Soviet Union. Hence the still popular portrait of von Braun as a visionary crusader, an opportunist who signed on with the Third Reich in a Faustian bargain only to follow his dreams.[19] Some of the pioneers clearly engaged in such imagery for self-protection, much like countless other Germans desperate to reshape their pre-1945 lives.[20] It is rather as though the inventors of the atomic bomb were to recall a youthful fascination with pyrotechnics, and claim that this is what they were pursuing all along. Why so

many American writers would abet this effort for so many years is ultimately a psychological issue tangled in the miasmas of the cold war,
beyond the scope of this book.

With this heavy caveat in mind, one could, like some space historians,
earmark the tumultuous year 1923, when hyperinflation sent the price of
a loaf of bread to 428 billion marks, sparking riots and plundering throughout Bavaria, Saxony, Thuringia, and elsewhere.[21] The German economy
and government were under assault from every angle, yet a twenty-nine-
year-old itinerant named Hermann Oberth self-published a ninety-two-
page book titled *The Rocket into Interplanetary Space* (*Die Rakete zu den
Planetenräumen*). Oberth was born in Transylvania, an Austro-Hungarian
province with an ethnic German populace that became part of Romania
after the war. Like myriad children around the world, he had been inspired
by the novels of Jules Verne, such as *From the Earth to the Moon* and *Around
the Moon*, and now claimed that his study of rockets dated back to 1907.[22]
After military service in the Austrian Army as a hospital medic he enrolled
in physics and astronomy courses at several universities in Germany, per-
severing to submit a doctoral thesis concerning space travel at Heidelberg,
where it was rejected in 1922.[23] That such an outré subject, presented as an
astronomy dissertation, advanced this far in a conservative academic
milieu can perhaps be explained by the nascent technics cult, but more
probably by the general chaos of the era.

When he published his book, Oberth was living in Munich. At that
time, the Bavarian city was the premier German hotbed for rabid right-
wing politics, especially among students. Its Nazi-instigated beer-cellar
brawls landed Hitler in the city's jail for a month in 1922. In the summer
of 1922, the same thuggish element spat upon visiting Reich President
Friedrich Ebert. Though the Nazi Party was banned in most German
states—especially after the gangster-style murder of Jewish foreign minister Walter Rathenau in June 1922 by former members of the pro-Kapp,
Hakenkruez-emblazoned "Ehrhardt brigade"—it remained open in
Bavaria.[24] In November 1923, Hitler staged his abortive "beer hall *Putsch*"
from the *Burgerbraukeller* in Munich, which faltered only when the local
army garrison sided against the uprising. While Oberth's name has never

been directly associated with any of the myriad factions there seeking to undermine or overthrow the Republic, the rocketry enthusiasts whom he influenced or worked with personally were never far removed from these circles. Interest in spaceflight often seemed to coincide with flight into hard-right politics.

Using his wife's household savings, he produced from his dissertation manuscript a volume that made three claims far beyond the scope either of its mathematical content or any engineering knowledge of the day: first, with present technology rockets could be built to climb higher than the earth's atmosphere; second, these rockets could carry passengers without health hazards; third, the technology could pay for itself "under certain economic conditions." This was all speculation for which Oberth had no more evidence than Verne, whose literary success must have looked attractive to an unemployed ex-student in a year of ruinous inflation, food shortages, France's invasion of the Ruhr, gun battles across Munich between rival political sects, and Hitler's *Putsch*. A book review celebrating Oberth's "Germanness" appeared in the right-wing *Deutsche Allgemeine Zeitung* newspaper in Berlin, demonstrating the friendliness of this political quarter toward quasi-scientific fantasies that struck many others as the stuff of pulp fiction.[25] Just how a small book on a bizarre subject by an obscure first-time author found its way into this particular Berlin newspaper is curious enough, but it is reasonable to assume that the link was ideological, the innuendo of "Germanness" pointing in only one direction.

Oberth might have sunk from further notice if he had not crossed paths with a budding Austrian showman also living by his wits in Munich, Max Valier. Just seven months younger than Oberth, Valier had served in an Austro-Hungarian aviation unit during the war and then turned to writing of a science-fictional flavor after quitting university physics studies. At the moment when Oberth was thinking of trying another, less technical book about spaceflight while devoting more time to earning a respectable living as a secondary school teacher, Valier sought his input to publish *Advance into Space* (*Der Vorstoss in den Weltenraum*) in 1924. With popular prose decorated by fanciful pictures of Verne-like spaceships, it enjoyed several printings and helped boost sales of Oberth's work. In lectures and magazine articles (e.g., "Berlin to New York in One Hour," "A Daring Trip to Mars") Valier continued to slake the public's apparent thirst for futuristic escapism with a naive—or perhaps cynical—disregard for reality. In 1939, Hitler said that he got to know Valier fairly well in Munich, calling him a dreamer.[26] For anyone with a fertile imagination, spaceships would have been an appealing subject compared to the hardships of everyday life in Germany.

Oberth and Valier soon parted ways, the former having less appetite for wild dreams such as rocket-powered transatlantic passenger planes. Valier, too, might have dropped from sight if he had not found a kindred spirit in Fritz von Opel, grandson of the automobile company founder. Opel had studied mechanical engineering at the Technical University of Darmstadt and then become director of testing for the family business, as well as its head of publicity. This was a fruitful combination of duties for him, leading to an interest in building stunt cars. Valier's constant search for income intersected with Opel's need to make advertising news, resulting in the construction of rocket-propelled racers that drew rapt attention from the popular press.[27] Their projects happened to coincide with a period of relative economic prosperity, if not normalcy, in 1927 and 1928—the so-called golden years of the Weimar Republic—that had not been experienced since before the war.[28] Moreover, 1927 was the year of Charles Lindbergh's sensational transatlantic solo flight between New York and Paris, a fabulous stunt that could only have underlined for many nationalistic Germans their humiliated condition. In 1927, too, some 130,000 Stahlhelm members in military garb took part in a Berlin march to show their loyalty to the old order, and the first overt Nazi anti-Jewish attacks were staged. In March, a gun battle broke out between Communists and Nazi Brownshirts on a train in Berlin-Lichtenfels—only the most notorious of a growing number of such clashes.[29]

In May of 1928, Opel performed a highly publicized demonstration of his "RAK 2" rocket car for several thousand Berliners at the AVUS (*Automobil Verkehrs und Übungs-Strasse*) racetrack through the Grunewald park in southwestern Berlin.[30] To reach speeds over 100 miles per hour, Valier acquired black-powder rockets from a company that had long manufactured them for whaling harpoons and naval signal flares. The rockets were simply attached in bundles to the rear end of a single-seat racing automobile and ignited. This spectacle spawned similar shows during the next year with ice sleds, railway cars, sailplanes, and bicycles, though Opel quickly moved Valier out of the spotlight.[31] The vehicles were of no practical value, since the solid-fuel rockets were an inefficient source of power and could not be throttled once ignited (RAK 2 careened from 0 to 100 mph in eight seconds). But they were undoubtedly crowd pleasers. For a brief time before the stock market crash of late 1929 brought another economic maelstrom, the imagined adventure of spaceflight was a subject of some popular interest in Germany, culminating in the Berlin premiere of director Fritz Lang's science-fictional movie *Frau im Mond* (*Woman in the Moon*).

This portentous movie did not come out of nowhere. Despite the presence during the 1920s of talented artistic directors like Lang, Berliners spent much of their precious entertainment money on low-brow productions cranked out by UFA (*Universum Film A.G.*), the dominant German film conglomerate. Many of these pictures were simple romantic comedies whose popularity stemmed from cultural glorification. Lang's famous *Metropolis* (1927), while cinematically ground-breaking, had nearly bankrupted UFA when its elaborate futuristic sets failed to boost a mundane script about the futility of resisting authority. In 1927, the company fell under the control of Alfred Hugenberg, former chairman of the board of the arms manufacturer Krupps and a Wolfgang Kapp associate, who had become an outspoken, radical, nationalist press baron.

In 1916, Hugenberg had teamed with shipping magnate Ludwig Klitzsch to produce propagandistic films defending the German war effort. UFA was formed in 1917 during the Ludendorff regime as a consolidation of the film industry under the Supreme High Command, bringing together business leaders from the most conservative circles of society, especially banking and the military.[32] Now Hugenberg desired movies that would instill what he saw as good German values, especially anything anti-Republic. In December 1926, the Reichstag had passed the controversial *Gesetz zur Bewahrung der Jugend vor Schund- und Schmutzschriften*

(Law to Protect Youth from Trashy and Dirty Writings), which established state censorship boards with the supposed aim of stemming the
popularity of low-brow, sometimes sexually explicit literature such as
detective, romance, and adventure stories.[33] Elected chairman in October
1928 of the far-right German National People's Party (*Deutschnationale
Volkspartei*, or DNVP), which constantly blamed the Republic for *Schund
und Schmutz*, Hugenberg proceeded to court the favor of Hitler and the
still marginal National Socialists in unconditional opposition to the Weimar government.[34] The DNVP, rooted in Junker circles of business and
state bureaucracy, had supported the Kapp putsch and was deeply anti-
Democratic, racist, and anti-Semitic. *Frau im Mond* was one of dozens of
dramas that suggested the immoral and dangerous side of cosmopolitan
life, especially in this case the corrupting influence of international finance
on pure native German ideas: a visionary but ridiculed professor is co-
opted by foreign investors to mine gold on the moon.[35] Lacking the cine-
matic brilliance of *Metropolis*, it depended on similar technical fantasies of
the current machine cult.

For the film's opening night at a UFA theater off Berlin's *Kurfürstendamm* commercial boulevard just two weeks before the Great Crash of
October 29, 1929, Lang and UFA struck a deal with Oberth to build and
launch what would today be called a liquid-fuel "sounding rocket," capable of rising into the stratosphere, or at least beyond the sight of sidewalk
gawkers, to publicize the story of journeying to the moon. This was beyond
Oberth's or anyone else's skills, however, and produced only financial and
emotional trauma for the writer, who left for Romania well before the
gala. *Frau im Mond* was neither an artistic nor a commercial success, but
its technical effects were prescient in several remarkable respects, including a "roll-out" of the moon rocket to its launch site from a huge assembly
building and the first numerical "countdown," which Lang devised for
dramatic tension. A publicity poster depicted a throbbing phallus of a
rocket roaring through outer space, but its basic composition—pointed
metal cylinder with fins around the fiery exhaust—proved to be realistic.

Though the movie and the fad quickly faded away as the German
economy disintegrated, the curious network of individuals drawn to per-

sonal involvement with rocketry continued to expand stepwise. Oberth had hired assistants to help him with the hapless movie job, each of the down-and-out type one might expect to answer an employment ad for such work during ruinous economic times. One, a wartime aviator and Stahlhelm member named Rudolf Nebel, who had also been a student in Munich after the war, was of the Max Valier ilk, possessing more energy as a pitchman than knowledge as a technician. Many years later, Oberth remembered him as a "great scoundrel" and "conman."[36] After Oberth's departure, Nebel took over the remnants of the failed *Frau im Mond* project and insinuated himself upon a group of science-fiction enthusiasts who called themselves—with or without humor can only be guessed—the Society for Space Travel (*Verein für Raumschiffahrt*, or *VfR*), which had been founded in 1927 partly under Valier's show-business impetus.[37]

Nebel tried to use his contacts and powers of persuasion to drum up money for building what Oberth had first envisioned for Lang. Eager to connect with anyone who would listen, he wrote to Adolph Hitler, Joseph Goebbels, and Hermann Goring in January 1930. These particular individuals were not just anybody, of course—what they stood for was by now well known. No funding came his way from the Nazi Party, however, which must have regarded him as even farther out on the fringes at this point.[38] Still, with financial devastation everywhere, he did find support from within the Reichswehr itself in the person of Lieutenant Colonel Karl Becker, an ultranationalist ordnance officer with an engineering doctorate from the Technical University of Berlin.

AT THIS MILESTONE, where military nurturance of rocketry begins, it is prudent to step back for a moment and question what relevance, if any, the activities of the spaceflight faddists actually possessed for the field's future development under weaponry auspices. Strictly speaking, their technical contributions were minimal. At best, they can be classed with myriad flying-machine seekers of pre-Wright vintage—some con artists, some obsessed tinkerers—who for whatever reason took a subject seriously that most people considered foolish, thereby creating a tiny seedbed for creative work, however ludicrous. The rocketry fad was very slight, however, compared to "Zeppelin fever" or the mass celebration of every new record set by German airplane pilots during the 1920s. Besides, it was based on completely imaginary technology. Except for Oberth, who at least understood some of the mathematics of ballistics and propulsion, the cast of players was dubious. That they melded into right-wing circles that ultimately connected with extremist military elements says much about the personal culture of Weimar technology at this grass-roots level. The psychological pull of rocketry was still far more potent than reasoned argument or concrete programs, a characteristic it shared with far-right German politics. The only thing in Berlin that was truly rocketing upward was unemployment, along with the radicalism of the anti-Republican right. What really held the rocketry circle together was joblessness.[39]

That one figure who would later achieve worldwide acclaim appeared in their midst around this time serves to lend them all more credibility than they otherwise deserved. At about the age of eighteen, for reasons that puzzled even his own father, Wernher von Braun joined Nebel's coterie of Berlin rocketry enthusiasts. His youth and social stature, at least, distinguished him from the gang of n'er-do-wells.

4

AN HEIR OF CREDIBILITY

Wernher von Braun was not the first teenaged son ever to mystify his father when it came to interests outside the home.[1] But the elder von Braun was in a better position than most to judge the company his boy was keeping. As a cultured family with direct personal connections to the political and economic events of their time, they were hardly ignorant about the murderous era taking shape around them. No sentient German beyond childhood could have been. Wernher's father had never been anything but a staunch defender of traditional, rigid Junker discipline. If there were any concern within the family about the young man's path at this age, he would have been firmly steered in other directions. To the contrary, every step he now took brought him closer to the National Socialist movement.

At about the same time that Rudolph Nebel was writing to the Nazi Party's leadership for financial support, Wernher became an active member of Nebel's rocketry group.[2] How the connection was initially made is unclear. Von Braun wrote many years later that when he "heard that the UFA film company had provided Oberth with a modest sum of money for his experiments, I offered him my services as an assistant."[3] On another occasion he wrote that he "had the good fortune to fall in with [Oberth] through the kind offices of Willy Ley," who had been the publicist for

Frau im Mond.[4] "As soon as I graduated from high school, Willy Ley, already a prolific popular writer on space and rocketry, introduced me to Professor Oberth," von Braun said.[5]

Neither Oberth nor Ley recalled such introductions, however. Ley remembered only that von Braun "visited me in my home one day, presumably that was during the summer of 1929," to discuss what subjects he should study in order to qualify someday as a researcher in rocketry.[6] He surmised that the teenager "must have appeared on the scene"—that is, joined the rocket club—"between March and May" of 1930. "Just how this happened I do not know. But I suppose that he may have visited Oberth the way he had visited me a year earlier." Yet Oberth himself retained no memory of how he had first met von Braun, other than that "he came to me."[7] His initial notice of him occurred not because he was any kind of wunderkind, but "through a clumsiness" during assembly work on a rocket project, Oberth recalled. "There are instruments to make holes in a stopper [rubber cork], but he used a hammer," Oberth remembered. "He was actually not very skillful with his hands."

Considering Wernher's tender age and lack of manual skills, the absence of any prior association with club members, plus the sheer social gulf between the baronial von Brauns and the *Lumpenproletariat* rocketry enthusiasts, it is curious that he could simply "fall in" with Oberth or Ley, let alone be so readily accepted into the group's hands-on shopwork. For certain, they needed money. The von Braun family was relatively well-off, so hope for some quid pro quo in this regard was likely. Having a Junker scion in their midst, no matter how wet behind the ears, could only enhance their credibility. It is also entirely possible that Wernher's father, insisting that he at least obtain sound educational advice before joining what could only have seemed to the elder von Braun a rather tawdry organization, might open a channel to Oberth or Ley through old associations with Alfred Hugenberg. It is inconceivable that Magnus, a former wartime press chief in the chancellor's office, would not have lines to Hugenberg. There is no record, only the vague, self-defensive recollections of decades later when "space travel" was the favorite German explanation for any personal interest in rockets prior to 1945.

A careful reading of these recollections reveals that they seldom connect with each other except in a most thematic fashion. In one published
account, supposedly based on the author's personal conversations with von
Braun, Oberth is portrayed as "installing an exhibit in a department store"
when he gets a call from von Braun. "Come right over," Oberth tells him.
The teenager helps set up the display and subsequently "attended the
exhibit and answered questions with wordy promises of travel to the Moon
at an early date."[8] While Oberth might have welcomed an extra hand for
such a job, his cautiousness would have precluded allowing a teenager to
act as any kind of spokesman, though the "von" in the boy's name might
have helped.

Even von Braun's personal opinions about *Frau im Mond* in later years
were contradictory, especially for someone who had clearly been influenced by the film's science-fictional imagery. He referred to *Frau im Mond*
in a 1956 memoir for the *Journal of the British Interplanetary Society* as "the
meretricious influence of the cinema at its most venal" produced by "grasping filmsters." Yet in a 1973 article for a Berlin newspaper, it was "Fritz
Lang's first-rate space film."[9]

All in all, these assorted memories suggest being framed to please their
particular audience, a not surprising habit for many Germans after World
War II. In fact, virtually all of the commentary about von Braun's youth,
whether from family members or acquaintances, is contained in statements made after the war, especially after he became an American celebrity. His own postwar reminiscences, which were often ghostwritten for
him, do not square well with documented fact, instead serving to paint a
blurry picture of a child infatuated with space travel to the exclusion of
anything else. "When I was 12 years of age, I had become fascinated by
the incredible speed records established by Max Valier and Fritz von
Opel," he recalled in 1963. He was twelve in 1924, but it would be another
four years until Opel's rocket car exhibitions took pace at the AVUS racetrack in Berlin. Records show that in 1925, von Braun transferred as a
student from the French Gymnasium in Charlottenburg to the Hermann
Lietz boarding school housed in Ettersburg Castle, near Weimar.[10] He
had apparently done poorly in mathematics and physics courses at the

Gymnasium.[11] He also recounted lightheartedly about how, around the same age, he had been collared by police for playing with pyrotechnic rockets tied to a wagon in his Tiergarten neighborhood. "I was released to the Minister of Agriculture (my father)," he explained, though his father did not become a cabinet official in the Weimar government until 1932.[12]

There is no doubt that the teenaged von Braun was fascinated by rockets, at least in the way that boys from industrial societies have been commonly enamored of trains, planes, and automobiles for generations. When he was confirmed into the Lutheran church at age thirteen, in 1925, he received a telescope from his parents as a present. He remembered sending away for a copy of Oberth's book after seeing an advertisement for it in an "astronomy magazine."[13] He joined the *Verein fur Raumschiffahrt* (Society for Space Travel) science-fiction club that circulated a magazine titled *Die Rakete* (The Rocket). He even tried his own hand at writing space-travel stories. There would be little reason to scrutinize his or others' memories of these youthful trivialities, if not for the romantic narrative of space exploration that eventually colored every inquiry into his adult life. Taken at face value, the recollections portray a grade-school and then college-aged student, living in a capital city where violent political and economic upheaval was polarizing the entire society, who was either perfectly insulated from these events or utterly indifferent to them. The first was impossible, even considering his rarefied social rank and removal to country boarding schools. To the extent that the second was plausible, it is ultimately more troubling.

AT THE END of March 1930, the relatively stable coalition government that had held national office since 1928 fell apart over the seething issue of unemployment.[14] With crucial midwifery from the army, the chancellorship passed from a Social Democrat to a right-wing Zentrum Party (Catholic) figure, Heinrich Bruning. This marked a fateful turn from Weimar democracy toward authoritarian regimes, culminating in Hitler's rise to power in January 1933. The Reichstag would not again congeal around a parliamentary majority, and rule by executive emergency decrees would become the norm. For the army, which had held itself together while tol-

erating the Republic for the past tumultuous decade, here was the opening
to skirt political control and take a more assertive position. Its goal was
abrogation of the Treaty of Versailles, rearmament, and the return of Germany to Great Power status on the world stage.

In general, older military leaders—epitomized by Paul von Hindenburg, elected president of the Republic in 1925—had remained rooted in
the Prussian monarchism of prewar years, while young officers responded
to the populist élan of Hitler's jackboot nationalism.[15] To ward off the
influence of various paramilitary brigades, as well as to maintain traditional discipline in the ranks, open relationships with extremist groups
had been discouraged since the abortive Kapp putsch, though Hindenburg himself was an honorary president of the Stahlhelm. While there
was no doubt that the military's sympathies lay on the far-right end of the
political spectrum in one fashion or another, the top command perceived
that it was in their overall interest to stay aloof from shrill organizations
like the Nazis, if only because they still depended on the Republican system for funding. In this regard they had been quite successful, keeping the
army's budget steady while other state institutions were cut.[16]

Rudolph Nebel and Lt. Col. Karl Becker thus met at an opportune
moment. One might posit that the meeting was made possible by the
army's budding confidence. Nebel and Becker were hardly peers in either
a social or intellectual sense, the former being a scurrilous character of
dubious education and the latter a career artillery officer with an engineering PhD. Only at a time when ultranationalist politics bridged such
gaping class divides could two such men find common ground.

Though the army had grown practiced at exploiting ways around the
Versailles limitations on manpower and weaponry, even testing forbidden
devices in the Soviet Union, serious testing of long-range rockets—albeit
technically legal under the treaty—beyond the amateur rocket club's
backyard toylike scale would have been out of the question while Germany was involved in fragile diplomatic negotiations to revise the pay-
ment of war reparations. The extreme nationalist right (and left) bitterly
opposed the Young Plan, named after an American chief of the General
Electric Company, whereby Germany received a pledge for immediate

removal of Allied occupation forces from the Rhineland, but agreed to continue annual payments of just over 2 billion marks until 1988.[17] Alfred Hugenberg persuaded Hitler to join a highly publicized campaign against the plan, which succeeded in staging a plebiscite on a law "against the slavery of the German people" that declared the Young Plan to be treasonous.[18] While the law was soundly defeated at the end of December 1929, the campaign gave the Nazis—still regarded as a lunatic fringe even by many anti-Republicans—valuable exposure on a relatively respectable political platform.

For a right-wing army officer trying to move forward toward rearmament in his area of specialization while avoiding the public minefield of these debates, *Frau im Mond* and Oberth provided the perfect cover of movie business hijinks. Becker agreed to give Nebel money from Army Ordnance to support Oberth in another attempt to build and launch his *Frau im Mond* invention. Nonetheless, it was a peculiar partnership, one that is difficult to analyze in retrospect regarding who was fooling whom. Nebel was without doubt a tenacious character and Becker was infected with the era's technological romanticism. They were both young and ambitious in a society that was unable to offer much to such a combination. Becker's relationship with Nebel could perhaps be compared with Fritz von Opel's connection to Max Valier—purely practical and easily dropped.[19]

ON APRIL 12, 1930, at the dawn of a decade that would propel Germany to its fateful pinnacle of power, Wernher von Braun enrolled as a student in mechanical engineering at the Royal Technical College of Berlin in Charlottenburg.[20] He had just graduated from a remote and primitive outpost of the Leitz school system on the North Sea island of Spiekeroog with mixed grades, finally scoring well in math and physics classes, but average to poor in everything else. Given his spotty academic performance and the offbeat reputation of such boarding schools, it is questionable whether he could have been accepted at a major non–*Technische Hochschulen* German university, regardless of family clout. He had distinguished himself, at least on isolated Spiekeroog, with interests in astronomy, and had been allowed to leave a year earlier than normal. The Technical College had an honored history in science and engineering, but was already a hive of Nazi activity, particularly among its students.[21]

In early May, he began a term of practical shop work at the Borsig Locomotive Factory in the Berlin suburb of Tegel, an apprentice experience required for all freshman engineering students. "During the entire summer of 1930, I left my parents' apartment, located in the Tiergarten section, at 5:30 A.M., and rode the streetcar through the endless Mueller-strasse to Tegel in order to punch my time card a few minutes before 7 A.M. at the entrance to the factory," he recalled many years later.[22] On his first day, an old master mechanic handed him a "piece of iron the size of a baby's head" and ordered him to file it by hand into a perfect cube. For six weeks, with only a gruff *Weitermachen!* (Go on!) from the stern foreman, he scraped and measured until he had an acceptable block "the size of a die." His "manly initiation at Borsig" taught him "the value of patience and perseverance," he claimed, making no wider observations about what must have seemed like an excruciating waste of time.

Thus did the Junker scion who was unskillful with his hands first meet the exigencies of the shop floor, at least in this folkloric memory. German industry was already staggering under the shocks of widening depression, so it is not surprising that he was given a functionally pointless task.[23] Borsig went bankrupt in the fall of 1931 and the government tried to refloat it with public funds. The company's owner, prominent Berlin industrialist Ernst von Borsig, was a longtime supporter of extreme right-ist organizations, including the Nazi Party, which he concealed out of concern for his firm's heavy dependence on contracts from the Republican state.[24] He had denied press reports in 1927 of subsidizing Hitler, but contributed personal money and raised funds among other Berlin business-men to help establish the city's NSDAP branch, believing the Nazis could lure workers away from Marxism.

This was not all von Braun was up to that summer while the nation roiled in a feverish and often violent election campaign. Every afternoon after putting in his hours at Borsig, he went to a nearby proving ground in Ploetzensee occupied by the Chemisch-Technische Reichsanstaldt, a

AN HEIR OF CREDIBILITY   53

government testing institute similar to the U.S. Bureau of Standards. Rudolph Nebel had finagled space there, probably with a good word from Lt. Col. Becker, to build and certify Hermann Oberth's rocket motor design. On the basis of this for-once concrete step forward and the army funds proffered by Becker, which Nebel recalled in his memoirs as 5000 marks, Oberth himself returned to Berlin for the summer. Given the dire state of the German economy and the hardship being experienced by people across the country, Nebel's figure—equivalent to about $1200 at a time when average per capita income was in the low hundreds—seems fantastic, but this would be in keeping with his reputation as a con artist. The ultimate goal was to launch Oberth's *Frau im Mond* rocket from the remote Baltic beaches.

According to von Braun's recollection, Nebel obtained nearly everything by barter. The immediate object of their attention might today be regarded as a school "science fair" project: a *Kegeldüse*, or cone jet. "It consisted of a greengrocer's scale bearing a pail of water from which protruded the tiny nozzle," von Braun wrote many years later. "A Dewar flask for oxygen, a bottle of nitrogen with pressure reducer, a petrol tank and copper piping to connect the latter and the Dewar flask to the motor completed the installation." Once the oxygen and gasoline started to mix at the top of the cone in the water bucket, someone tossed a flaming rag at the Kegeldüse to ignite it and then dove for cover. By watching how much the scale was deflected, the motor's thrust could be measured. "Professor Oberth supervised the proceedings with tight-lipped saturninity," von Braun recalled. "The latter manifested itself in sarcastic remarks delivered in a Transylvanian brand of German which fell heavily on the ears of his listeners."[25]

On July 23, in the presence of an official from the Reichsanstaldt, the group fired their little Kegeldüse for ninety seconds and generated a thrust of 7 kilograms. This was considered such a milestone that a photo was taken of the group, which shows the eighteen-year-old Wernher neatly attired in jacket, neck tie, and knee britches, grasping what may be the unphotogenic cone nozzle itself while the older men—Nebel, Oberth, Klaus Riedel, and others—pose beside a finned rocket model of some 6 or 7 feet tall that was probably a cheap prop from *Frau im Mond*. Nebel stares warily toward the photographer, always eager to get a good publicity shot. Oberth stands next to the symbol of his long frustration. Riedel, the group's "chief designer," studies an old-fashioned signal flare, essentially a big bottle rocket, that looks impressive but had nothing to do with their work. Von Braun appears to be listening intently to what Oberth is telling a rotund man in a fedora, possibly the Reichsanstaldt official. The teenaged Junker is holding onto reality here, at least. If nothing else, the Kegeldüse trial demonstrated how far away a Baltic launch was in every sense.

Oberth soon went back to life in Transylvania, again leaving the motley Germans to their fates. Oddball though he was, he did not trust Nebel and saw no future in the group's tinkering. Perhaps he alone understood the gap between their fantasies and the tedious, systematic steps forward that are the reality of technological development. Nebel and the rest clearly had nothing better to do with themselves. Von Braun was just a student, albeit from the aristocracy, who could only have found the older men to be very peculiar types, given his usual social milieu. It strains the imagination to believe that his parents knew exactly what he was doing every day after leaving Borsig.

As soon as Oberth was gone, Nebel took control of the group, moving their experiments to an abandoned ammunition storage depot in the run-down Berlin suburb of Reinickendorf, site of present-day Berlin-Tegel International Airport. He and several other men began to live crudely there, too, which was probably a reflection of their desperate economic straits rather than sacrificial devotion to technological progress like the Wright brothers at Kitty Hawk. This would have been impossible without the military pulling strings for him, though von Braun later recalled only that Nebel "persuaded the city fathers to grant us a lease on it—free and for an indefinite period."[26] They put their flammable gear in an old blockhouse and called the place *Raketenflugplatz Berlin*—a German meshing of nouns that in English comes close to "Berlin Rocketport"—an absurd name that Nebel would have possessed the flare to embrace. "We had no financial backers," von Braun said, either dissembling or forgetting about the army's support, which he might have been unaware of at the time. It is safe to believe that his father, at least, would have recognized behind-the-scenes help from that quarter in occupying the ammo dump land and bunkers.

The move came two weeks after a political earthquake, the portentous September 14 Reichstag election, in which the radical right saw astonishing gains as Hitler's party grew from 12 to 107 deputies on a popular vote of almost 6.5 million.[27] The number of jobless Germans was well beyond the 3-million mark (out of a total population of 65 million), with about 2 million receiving unemployment insurance benefits, which would push the system toward bankruptcy. Tens of thousands of people had attended Hitler's demagogic speeches of national redemption around the country during the frenetic election campaign, including at least 16,000 at Berlin's Sportpalast on September 10. His name was now in the air everywhere. When the new Reichstag convened in mid-October, Nazi deputies arrived in their brown uniforms and marched into the hall in military formation, repeatedly causing suspension of the proceedings. While achieving little of substance with such obstructionist tactics, they did manage to force the government to ban showing of the American film version of Erich Remarque's antiwar *All Quiet on the Western Front*. In the center of Berlin, brown-shirted gangs vandalized Jewish shops.

Years later, von Braun conflated the *Frau im Mond* period with the Kegeldüse summer, making it seem as though UFA were still funding Oberth at Ploetzensee in hopes of obtaining a publicity rocket for the movie.[28] Nebel put von Braun's clean good looks and aristocratic manners to work whenever he was around, taking him along on "scrounging" trips to collect scrap materials from local businesses or letting him speak to curious Raketenflugplatz visitors.[29] This handsome Junker engineering student was an infinitely finer public face for Nebel than any of the unemployed characters baching it on soup-kitchen handouts at the ammo dump.[30] Von Braun would play this role at ever higher levels for the rest of his life.

5

CHILDHOOD'S END

Y THE BEGINNING of 1932, with the Great Slump coming less than ten years after the Great Inflation, more than 6 million Germans were out of work, a catastrophic rate that would hover near one-third of the entire labor force. For unionized workers the figure was even higher, near 45 percent. The jobless and their dependents made up at least one-fifth of the population. Income of those who had any was almost 30 percent lower than before the Great War. The government, barely afloat as a parliamentary democracy on stormy seas since 1918, was sinking, and 1932 would see the denouement when all worth saving was lost. This was not a time of normality anywhere in Germany—with unemployment at such levels, every corner of life was wracked by crisis. It has become clear to historians that the mixture of humiliation from defeat in the war and desperation from economic collapse delivered proud Germany into the hands of the monstrous side of human nature (though the abysmal depth of depravity to which she would slide will always defy concise explanation). As Eric Hobsbawm recalled seventy-five years later, few Germans had really wanted the Weimar Republic and most accepted it at best as a *faute de mieux*.[1] By the end of the year, it would metamorphose into an entirely different, sinister vessel.

Though the government of Chancellor Heinrich Bruning—an austere

conservative nationalist who had been ruling by dictatorial, albeit constitutional, emergency decrees since 1930—made savage budget cuts in the
face of vertiginous depression (Keynesian remedies such as deficit spending had not yet been devised), military funding did not suffer as much as
other categories. His curtailment of social programs and the civil service
exacerbated hatred of the Republic. Teachers' salaries and unemployment
benefits were slashed, for example, but Bruning found money to build a
new battleship, which was permitted under Versailles treaty restrictions
because it had no strategic importance. American capital had helped to
fuel an economic revival during the late twenties; now, because of a flood
of foreign credit withdrawals from banks, growing loss of domestic
accounts, and major bank failures during the summer of 1931, Germany
no longer possessed a free economy. The Reichsmark could not even be
exchanged for foreign currency. Radicalization of the German people
surged upward all the while, with the Republic serving as a catch-all
demon along with the Jews.[2]

German youth of every class were especially pummeled by the economic maelstrom, with little hope for finding work, let alone starting
careers. Unemployment for men between ages eighteen and thirty was far
above the average. Whatever idealism survived in these grim times was
fodder for the political extremes. "Large parts of middle-class youth no
longer think in bourgeois terms," observed a typical newspaper editorial.
"They think either Marxist or fascist."[3] In 1932, Hitler was only forty-
three, Goring thirty-nine, and Goebbels thirty-five. No matter how brutal
and unscrupulous their vitality, they stood in stark generational contrast
to calcified old leaders like Paul von Hindenburg, eighty-five, who had
been voted *Reichspräsident* largely on a wave of reactionary nostalgia after
the death of Social Democrat Friedrich Ebert.[4] No less than 43 percent of
the 720,000 new members who joined the Nazi Party between 1930 and
1933 were under thirty years old.[5]

Since their first Reich Party rally at Nuremberg in 1927, the Nazis had
perfected how to manipulate the hopes and hatreds of masses of resentful
people, thereby growing from a peculiar splinter group into a national
movement.[6] The party benefited not only from economic crisis, but from

the affinity of conservatives in the army and civil service, as well as the agrarian and industrial lobbies, all fueled by ubiquitous anti-Weimar vitriol. Hitler promised something for anyone who was alienated or ambitious, no matter how vague and contradictory, so that eventually large percentages of people supported him, or at least did not actively oppose him. In this light, perhaps it begins to be possible to understand how individuals who were swept along by the Nazi tide could many years later deny knowledge of its dark side or claim personal motivations—like interest in space travel—that sound absurd under the circumstances. Whatever their dream, the Nazi vision of the future encompassed it. In July 1931, the national organization of the General Students' Unions had been taken over by the National Socialist German Students' League (*Nationalsozialistischen Deutschen Studentenbund*), led by twenty-four-year-old Baldur von Shirach, who had been raised like Wernher von Braun in an aristocratic family of wealth and culture and sent to a country boarding school directed by an associate of Hermann Lietz's.[7] Young men from this high station in life needed nothing besides a future, and the Nazis were more and more in a unique position to offer them one, despite being widely regarded as vulgar by the upper classes.

The most broadly seductive feature of Nazi doctrine, which was molded from many existing lines of thought in German politics and shaped according to hodgepodge opportunism, never forming a consistent ideology (even nationalism and socialism being somehow congruent), was *Volksgemeinschaft* togetherness within a racial German community. The attendant fanatical nationalism and superhuman führer personality cult were of narrower direct appeal, though the latter was everywhere a potent organizing force. Anti-Semitism was a fervent feature of the Nazi campaign—there was no generous future for German Jews—but of relatively lesser significance at this point, certainly never a unique feature of the NSDAP. Its genocidal centrality to Hitler's thinking should have been obvious to anyone who read his best-selling autobiography, *Mein Kampf*, but actual state-sponsored mass murder was inconceivable to a healthy mind. Despite not holding German citizenship until February 1932, Hitler was a master tactical organizer and rhetorician rather than ideologist, whose demagogy united broad existing currents of anti-Semitism and anti-Communism. Germans wanted to feel good about being German— a hardly unusual desire among nations, especially defeated ones, but perilous in the here and now to anyone considered un-German.

The Nazi appetite for deliberately provocative violence was in plain sight every day on Berlin's streets, impossible to ignore. In brawls with members of the Communist *Rotfront* (Red Front, almost entirely a party of the unemployed) or Social Democratic *Reichsbanner*, elite SS corps in sharply tailored black uniforms joined the tawdry brown-shirted gangs, which chanted verses such as "Song of the Storm Columns" as they marched:

So stand the Storm Columns, for racial fight prepared.
Only when Jews bleed, are we liberated.
No more negotiation; it's no help, not even slight:
Beside our Adolph Hitler we're courageous in a fight.[8]

No sentient Berliner could be blind to these events. Ten years earlier, after the murder of Walther Rathenau in 1922, the Reichstag had passed a law making subversion of the Republic by word or deed a treasonable crime, but a right-leaning judiciary aimed it mostly against the Communists.[9] In Bavaria, always a reactionary bellwether, the state government officially supported radicalism on the right.

During this bitter time, the unemployed men sheltered in the bunkers at the Raketenflugplatz—the irony of the name grows heavier the more one understands the dire earthbound straits of the surrounding society— started calling the Junker college student who appeared from time to time "Sonny Boy." No doubt piling on sarcasm, they took the moniker from the international hit tune of Al Jolson's 1928 film *The Singing Fool*. It was a puerile song crooned in front of a Christmas tree to the cutie-pie son of Jolson's hard-knocks protagonist, who loved his little boy who had been sent from heaven to make gray skies blue.

The men could only have been painfully aware that Wernher spent most of his time in a pampered stratum at posh home addresses far above the daily mayhem of street-level Berlin, let alone their slummy abode. He had a family allowance in his pocket, took flying lessons at the renowned Grunau glider center in Silesia, and spent holidays at Oberwiesenthal, a Silesian estate—that is, in the Junker manner, a large farm and dependant village—that his parents had purchased as a second home in 1930, a year of widespread compulsory auctions of bankrupt farmland.[10] Silesia was the traditional turf of Germany's super rich—the term "Prussian magnate" implied Silesian, while the country's poorest population did the actual work there. The von Brauns had more than just survived the postwar years, and their oldest son was enjoying the kind of privilege that would have been available to him had the Great War and its tenebrous aftermath never happened. The Raketenflugplatz and "Sonny Boy" existed in the same dimension of unreality.

Wernher had also passed the spring and summer months of the previous year taking courses at the Swiss Institute of Technology in Zurich and then motoring to Greece with a rich American medical student he met there named Constantine "Ntino" Generales. His reasons for leaving the Technical College in Berlin, only to return after the lark with his new friend, were never explained.[11] But there had been so much brawling between Nazis and Communists in Berlin since November 1930 that the college, which was a center for National Socialist agitation, had to close temporarily in June 1931. When von Schirach's Nazi student league took over the general student organization in 1931, they won 66 percent of the vote at the Technical College, the highest in the city.[12] Regardless of whether von Braun was involved in these matters, he could not have been unaware of them. At the very least, the interlude abroad was evidence of a lavish degree of mobility unavailable to most Germans struggling just to make ends meet.

One macabre vignette from the first months of their friendship holds interest in the light of future events. Many years later, Generales recounted how Wernher showed him a letter he had received from Albert Einstein that was "replete with mathematics and referred to rocket propulsion."[13] Though the letter, if authentic, has not survived in any archive, Generales recalled being so impressed that he suggested they test the rigors of spaceflight for themselves. They proceeded to attach live mice stolen from a biology lab onto the rim of a wheel from Generales's bicycle and spin it

until the rodents were torn apart by centrifugal force, which they estimated to be as high as 220 g. "Generales then dissected the victims in
search of physiological data," von Braun later wrote. "He was able to state
that cerebral hemorrhage was prominent among lethal effects suffered
under high accelerations."[14] Generales remembered that "all the organs in
the chest and abdominal cavities, as well as the brain, were displaced and
torn in varying degrees from the surrounding tissue."[15]

These ridiculous, ghoulish experiments, pseudoscientific and completely divorced from academia, which both men later cited as the birth of
"space medicine," were halted by a landlady's "violent objections" to blood
stains on the walls of Wernher's room.[16] According to Generales, she
"became infuriated; seized my notes as evidence of nonsensical cruelty
and torture; and threatened to evict us and notify the police." They then
"had no choice but comply with our nonscientific but meticulous landlady." Generales, a twenty-four-year-old Harvard graduate who was supposedly splicing together his medical education as he hopped among
various university towns between Athens and Paris, purported that he
later resumed this investigation "with a larger centrifuge" at the Sorbonne
with the help of a lab technician identified, rather like Gertrude Stein's
maid, only as "Helene."[17]

The episode could be dismissed as silliness if von Braun and Generales
had been little boys, but they were young men. The equation of the landlady's outrage with a "nonscientific" attitude was small but pointed evidence of how moral objections can be belittled as antiprogress, a large
issue in the history of twentieth-century science and technology, one that
weighs heavily upon von Braun's career during the Third Reich. When
they returned to Berlin at the end of the summer of 1931, Generales fell in
with the Raketenflugplatz scene, lending his car for the group's use.

During their absence, Rudolph Nebel and Klaus Riedel had put
together what von Braun later remembered as a "hasty design" for a small
rocket that Nebel dubbed the *Minimum Rakete*, shortened to "Mirak."[18]
Nebel displayed some levity by quipping that this also stood for "minimum effort plus miraculous achievement." Taking advantage of a load of
thin aluminum pipes that Nebel had scrounged, Mirak consisted of an

operational version of the Kegeldüse mounted at the leading end of one of the pipes.[19] If everything went well, this "nose-drive" device rose from launching rails to an altitude of about a thousand feet, and then fell back to earth on a parachute packed in the other end of the pipe. They adopted the more dramatic name *Repulsor* from an old German science-fiction novel and Nebel tried to charge spectators one mark apiece.[20]

Nebel also hawked copies of *Raketenflug* (Rocket Flight), a self-produced pamphlet about the history and futuristic promise of rockets, prominently featuring himself as founder and leader of the *Raketenflugplatzes* (plural) *Berlin*. A full-page photograph showed him in clean white lab smock, busily speaking on a telephone—symbol of trendy affluence—with a *Frau im Mond* poster behind his head like a window into outer space.[21] This hucksterism can be compared to itinerant "Professor Marvel" medicine shows of Depression-era America, which plied poor rural roads as depicted in the 1939 film *The Wizard of Oz*. Like the Nazi Party, to whose leader he would soon be writing with *Sieg-Heil!* over his signature, Nebel knew that desperate people would turn almost anywhere for relief.

Along this picaresque, innocuous line the narrative of German rocketry might have continued indefinitely. Nebel's band had built a somewhat workable toy, as had one or two other groups (not counting Robert Goddard's pioneering experiments in the United States).[22] Even a reliable *Repulsor* would have been of no use or interest outside the realm of hobbyists. But Nebel's bunkum had reached into military circles, eased by right-wing affinities. Something in the era's Zeitgeist maintained the army's attention— at least the attention of a few young artillery officers, most of whose colleagues would have found rocketry to be nothing more than amusing at this point. Whether due to sensationalistic press reports, a desire to control anything with potential for weapons development, or the reactionary modernism of Nazi identification with futuristic technology, they would keep dipping into the circle of science-fiction enthusiasts until they found something useful.[23] It turned out to be just one particular individual.

AS THE DYSFUNCTIONAL economy plunged toward its lowest level in the spring and summer of 1932, Weimar government blundered toward Hitler's ascension to power. The launching of toy rockets at the Raketenflug-platz took place during tumultuous presidential elections that attracted greater popular participation than ever before in Germany. In January 1932, Bruning—who was being called the "Hunger Chancellor" because of how his cutbacks were strangling the economy even further—held negotiations with Hitler and Alfred Hugenberg to see if the rightist par-ties would support extending Hindenburg's presidency without the national elections that were due in March, pending favorable outcome of diplomatic conferences on reparations and rearmament.[24] Both refused and an especially violent, irrational campaign ensued.[25] In the election on March 13 and the runoff that followed on April 10, the senile field mar-shal won the largest share of votes (49.6 percent and 53 percent, respec-tively, to Hitler's 30 percent and 37 percent), though many who voted for him were anti-Republic. During the runoff campaign, Hitler amazed the electorate by flying in an airplane from town to town to give speeches across Germany. Goebbels's brilliance in staging spectacles that equated his candidate with a glorious future, while additional propaganda zeroed in on specific groups of voters, relegated competitors to lame imitation.

Against the urging for a coup d'état by Ernst Röhm, commander of the nearly half-million-strong storm troopers, Hitler sensed that momentum was on his side and insisted on pursuing a legal path to power via upcom-ing elections at the state level. The Bruning government then proceeded to ban, on April 13, all Nazi paramilitary organizations. Their roughshod ubiquity was helping to scare off foreign investment, but even the Wehr-macht was nervous about their numbers, despite wanting them as auxil-iary troops. As Hindenburg, the embodiment of Prussian conservatism, felt increasingly outflanked on the right, his working relationship with Bruning soured, weakened additionally by intrigues among army factions and various economic interests as it became more and more clear that Hitler's rising strength demanded more than lip service.

At the end of May, after Hindenburg sought repeal of the paramilitary ban and dissolution of the current Reichstag, Bruning resigned, effectively ending Germany's dozen-year experience with parliamentary democracy. From this point until Hitler climbed the steps of the presidential palace seven months hence, control over Germany's future teetered between those who desired a return to Wilhelmine monarchic order and those who wanted a dictatorship. Democracy fell off the agenda as the parties that might have supported it dithered away their opportunities and splintered apart. The only opposition of significant ardor was Communist, whose membership never made inroads with the army or beyond the battered working classes. Hindenburg replaced Bruning with a stone from the only wall still standing that appeared to give the old order some degree of influence over the nation's course. Franz von Papen, a relatively unknown nobleman and monarchist, had neither popular support nor any stake in the Reichstag.[26] Essentially a surrogate for men who had lost touch with political reality, he applied an old-time conservative-authoritarian hand to the levers of government during his few months in power.

The cabinet he appointed consisted of similarly stationed Herren and became known derisively as the "cabinet of barons." Of the ten members, seven were noblemen, two were directors of major industrial firms (Krupp and I. G. Farben), and one—Franz Gürtner—was a Bavarian minister of justice who had done favors for Hitler since the 1923 Munich putsch.[27] None was a deputy in parliament. The new minister of agriculture turned out to be of special interest at the Raketenflugplatz. He was Magnus Frieherr von Braun, young Wernher's father.

"Lieber Braun, wollen Sie mit mir ein Kabinett von Gentlemen bilden und das Reichsernahrungsministerium ubernehmen?" (Dear Braun, do you want to form a cabinet of gentlemen with me and take on the Ministry of Food?) Papen asked him with anachronistic and somewhat anglophilic charm.[28] They had known each other in government circles for many years and Papen had often been a guest in the von Braun home. A seething crisis in German agriculture, exacerbated by conflict over a plan in the Bruning government to settle farm workers on bankrupt estates (which Hindenburg blasted as "agrarian Bolshevism") and eliminate subsidies to Prussian estate owners (which included Hindenburg himself), had played a significant part in Bruning's demise, so this was not a trifling appointment. Magnus possessed by birth the proper social credentials, of course, and shared the political bona fides of anti-Democratic disdain for the Reichstag ("Deutsche Luftreederei," German hot air, he called it), as well as anti-labor, pro-landowner economic views.[29] He was also the choice of the largest and most influential farm organization, the *Reichslandbund*, or Rural League, which had openly supported Hitler in the April balloting. It was the Rural League's president, Count Eberhard Kalkreuth-Nieder-Siegersdorf—referred to as "our friend" by the architect of Nazi agricultural programs, Walter Darré—who first informed him about the availability of the cabinet post. In June 1932, Papen became the first chancellor to attend a convention of local chambers of agriculture (*Deutsche Landwirtschaftsrat*), where he spoke of "Christian-national and social, not class and international" values—the code language of anti-Semitism.[30]

Before 1930, agriculture had been of little interest to the Nazis, whose roots were urban.[31] Early Socialistic ideas about land reform had sounded hostile to both farmers and estate owners. In the late-1920s, however, Hitler began to apply his opportunistic editing to these issues in search of political support, while the deepening depression in rural areas offered a new source of votes. The racist *Blut und Boden* (blood and soil) ideology was the result. Hindenburg, as an East Elbian estate owner himself, listened to the problems of agricultural interest groups, so gradual infiltration of farm organizations was an especially fruitful tactic for the Nazis. By the time Magnus von Braun was tapped to be Papen's minister, the party had established under Darré's direction a decisive grass-roots network among farmers and landowners all the way up to the venerable Rural League. Like many aristocrats, von Braun mostly regarded the Nazis as uncouth and was no match for their ruthlessness. The time was not yet ripe for Darré to be named minister of agriculture—that would come in a year—but by being socially acceptable to Hindenburg and politically acceptable to Kalkreuth, von Braun was a perfect placeholder.

While the nation was riveted by these events, Rudolph Nebel continued his dogged attempts to get more cash from the military. His instinct for fundraising was good, if not his technical sense, since there was no other potential source. Although Wernher von Braun turned twenty on March 12, and therefore could have voted in both the March and April elections that would clearly have been of keen interest to his father, there is no record or personal recollection of his having done so. All that is known is that he spent several weeks after his birthday completing his glider lessons.

Around the same time in April when the paramilitary ban was announced, several officers from Army Ordnance paid a visit to the Raketenflugplatz. For a group that shunned publicity as much as Nebel sought it, these appearances necessarily took place in mufti. "It was with no particular surprise that we recognized three inconspicuous civilian visitors who called on us as representatives of the Ordnance Department," Wernher wrote many years later.[32] They were Becker and two assistants, Captain Walter Dornberger and Captain Ernst von Horstig (whose name carried the honorific "Ritter" [knight], as von Braun's carried "Freiherr" [baron]). The "unobtrusive but sagacious visitors" showed no interest in Mirak I, von Braun recalled. Instead, they asked about technical matters such as thrust balance and wanted to see whatever diagrams and test data the group possessed. They also showed some curiosity about the untried Mirak II, a bigger version of Mirak I. Three veteran artillery experts could quickly assess the gap between Nebel's overactive imagination and reality, while strutting their professional superiority on technical details. And they had no patience for space-travel fantasies.[33]

Mirak II was a bit more than a toy, however, and perhaps not some-thing the army would want in the hands of already well-armed paramili-tary organizations, such as the Stahlhelm, of which Nebel was a member.[34] It was about a dozen feet long, weighed around 45 pounds at takeoff, and Nebel outlandishly claimed it could rise as high as 5 miles. There were obvious public safety issues, at the very least, even if its range was only a fraction of this. Its long, thin profile (only inches in diameter for most of its length) would have raised intuitive questions in an educated observer about whether it could maintain so-called arrow stability during flight. Becker thus made a prudent offer, more like a bet, that if the Raketenflug-platz men could successfully launch Mirak II from the army's proving grounds at Kummersdorf, some 25 miles south of Berlin, then he would cover their expenses. It took more than two months for them to put some-thing together, during which time the father of their youngest member became a powerful *Reichsminister*.

Several descriptions exist of what took place near the end of June at Kummersdorf, each composed decades later except for a brief army technical report. They paint a romantic picture of two private cars from the Raketenflugplatz, one carrying the spindly Mirak II on its roof and the other fuel and tools, driving through predawn streets of Berlin toward a hush-hush rendezvous with Captain Dornberger at the proving grounds. Space historians have posited that the army was trying to maintain strict secrecy, never considering that Berlin was in the middle of the worst period of civil violence in its history, that the sight of a car with a weapon-like object on its roof might have been more alarming to police or anyone else at night than in broad daylight. It seems likely that Becker did not seriously care whether Mirak II made it all the way to Kummersdorf or not, since the odds of a successful launch were miniscule. If by some chance the rocket arrived and performed as Nebel promised, then he would have kept it within military confines. If not, then nothing of any value was lost.

Nebel, Klaus Riedel, and von Braun—whom Nebel surely brought along to take advantage of his new celebrity—made it to Kummersdorf without incident, but Mirak II rose only for a moment from an artillery firing range before swerving horizontally and crash-landing. "The ordnance experts announced that Mirak II was far too unpredictable to meet their requirements," von Braun remembered long afterward. "Even Nebel's irrepressible optimism failed to induce them to carry on."[35] Becker had done Nebel one last favor, and now he was done with him. There would be no more commerce with the Raketenflugplatz.[36]

Only in the light of future events was the abortive launch notable, because it marked the beginning of young von Braun's direct personal association with the German army. Until then, the naivete of a teenager enthralled by science-fictional tales about space travel was perhaps sufficient to account for his presence at the Raketenflugplatz, setting aside the social incongruities. But the army was not neutral territory in any way and von Braun was not a simple tyke from the hinterlands. Since the debacle of the Great War and the political tumult of its aftermath, the army had been synonymous with extreme right-wing nationalism and revenge against the Versailles diktat. This was its reputation and raison d'être, which his father knew as well as anyone, and any thought that it could be otherwise would have been delusional.[37] Dornberger's memoirs make it clear that there was never any other purpose in the army rocket program besides building weapons.

Nebel tried in vain to rekindle his relationship with Army Ordnance. Only the son of the new minister of agriculture was warmly welcomed by Becker, however, whom von Braun found to be "by no means such an ogre" as Nebel had made him out to be.[38] The threadbare Raketenflug-platz was now of no importance to the artillery officers, except perhaps as something to suppress for the sake of secrecy. There are various recollections, all composed decades after the fact, of von Braun's next steps, in which he tried to bring his friends with him into the army fold, but they are not credible. Nebel, a war veteran, is remembered as having been "anything but enthusiastic about submitting again to the rigors of military control," yet he had spent years soliciting Becker and knew perfectly well that military money came with strings attached.[39] Becker would never have tolerated him on the inside, anyway. Riedel supposedly believed that private enterprise could finance space travel, a patent absurdity.[40] The only purpose these stories would serve would be to build the myth that the Raketenflugplatz possessed substantive technical importance, which it did not. Becker wanted von Braun, without whom the Raketenflugplatz would be of very minor historical note.

If perchance von Braun did actually try to convince them to come with him, a more plausible explanation is that he still did not possess the manual skills required to construct even a Mirak. It was his "theoretical knowledge"—that is, his ability to perform some of the mathematical calculations in which artillery specialists take pride—that had impressed Captain Dornberger, for example.[41] He was, after all, a college engineering student who had spent a few weeks filing an iron cube at Borsigwerke.

The army's focused interest in von Braun may have been vetted with his family around this time during a social encounter between his father and Captain von Horstig at a Reichswehr headquarters dinner.[42] The class bond between the baron and the knight would have sufficed to protect the young scion. In any case, favorable arrangements did not take long to make. Having finished only half of his undergraduate studies at the Technical College, von Braun soon found himself enrolled as a doctoral candidate in physics at the august Friedrich-Wilhelm University, with a paid contract from Ordnance to help develop liquid-fuel rocket motors at Kummersdorf.[43] Becker already had in place a secret military research program for select graduate students. Germany had pioneered the university-military research partnership that produced chemical weapons and synthetic ammonia during the Great War, and von Braun, only twenty years old, was being groomed at top speed to flesh out the generation of youth that Versailles had prevented the Reichswehr from recruiting. As in ages past, his Junker pedigree was as coveted as his facility with equations—much more, in fact. Less than two years after starting college, he was attracted by the prospect of pursuing cutting-edge research for the army under the highest academic imprimatur. This was the kind of command over the future that the Weimar Republic had never managed and the Nazis now stood for in toto.

Nebel, too, knew that the young von Braun was a valuable commodity, but the best he could do was quickly place him on the board of the Society for Space Travel, which in 1932 must have sounded even more odd than in 1927. The rather self-important directors balked, apparently sensitive about the effects on their image of allowing a twenty-year-old into their ranks, but were prevailed upon to elect him. It was an empty gesture, however, as von Braun became "less accessible," as one member put it, while his father's prominence and the Reichswehr took over.[44] There would be no more days of tinkering at the Berlin Rocketport. He soon resigned from the club and left it all behind in childhood.

"To us, Hitler was still only a pompous fool with a Charlie Chaplin moustache," he wrote, decades later, about this moment in history.[45]

6

"FINGERS IN THE PIE"

THE REMNANTS OF *Konzentrationslager* Mittelbau-Dora, near the Harz Mountain town of Nordhausen about 125 miles southwest of Berlin, consist today of brick foundations from demolished barracks, a restored crematorium, a bunker that served as an inmate prison-within-a-prison, sections of fence with rusted barbed wire, the old muster ground, and grassy fields that suggest the scope of the original 800-acre compound. Two gaping holes wide enough to accommodate railcars in an adjacent hill called the Kohnstein are the entrances to an immense underground tunnel complex with some 7.5 miles of corridors comprising almost 35 million cubic feet of space. The camp and the tunnels form a single locale, where starting in late 1943 slave laborers mass-produced in the world's largest subterranean factory the *Vergeltungswaffe*, or "vengeance weapon," known as the V-2 rocket.[1]

In the gory constellation of German concentration camps, Dora never received the world's attention to the degree of Auschwitz, though ranking these charnel houses on any scale of terror would be a futile exercise. Dora was initially a satellite of the Buchenwald camp, about 40 miles away near Weimar. But though some 20,000 prisoners died at Dora, it was never an extermination camp, per se, had no gas chambers, held no prisoners identified solely as Jews until the summer of 1944 (and then relatively few), and from 1945 until the reunification of Germany in 1990 was practically impossible for Westerners to visit because it was in the Soviet zone. Despite what was revealed about atrocities at Dora during the Nuremberg War Crime Trials in 1947, the transcript was classified by the U.S. Army until 1981. Former French prisoners called Dora "le camp oublié."[2]

All of these factors served to throw a blanket of inattention and ignorance over Dora for five decades, allowing statements from former overlords there to go unexamined. For example, to say that one had been at the rocket factory but never at the concentration camp during the war would be an obvious lie to anyone who could see for oneself that these were the same place, regardless of whether a fence stood between two sectors.

In the ten years that had passed since Wernher von Braun joined the German Army's fetal program to develop serviceable liquid-propelled rockets at the end of 1932, the Third Reich had fulfilled Hitler's goal of returning the country to Great Power status. The war he launched in 1939 to secure his dream of a millennium of German dominance now gripped much of the world. His "Germania" was not just a military juggernaut, but a totalitarian state that enforced racial hierarchies and political conformity with pervasive terror. This heinous side of the rejuvenated nation had always been part of his rhetoric and was now condoned, directly or indirectly, by millions of Germans.

But the time of blitzkrieg and easy conquest was already past.[3] At the end of January 1943, after hellish winter combat in Stalingrad, the starving fragments of the Sixth Army of 300,000 men became captives of Stalin, ending the ability to conduct major offensives. That same month, vast formations of American bombers began daytime raids over German cities. The war was essentially lost, which old-line Prussian generals comprehended, and the remaining time could be a salvage operation at best. As commander in chief of the army, however, Hitler was now steering it as he had always run the Nazi Party, concerned foremost with propagandistic effect and not bothered by factual reality. Wishful thinking thus took control of mortally wounded legions, fueled by the brutal dogma that National Socialist fanaticism could make up for inferiority in numbers. As his distrust of the Wehrmacht generals rose, more power accrued to the SS.

This was the atmosphere in which production of long-range ballistic missiles took place at Mittelbau-Dora. They were an expression of desperation, not confidence. As such, their manufacture turned into a technological grotesquerie in which thousands of human lives were consumed under sadistic circumstances to obtain a device that was no more than a deadly nuisance for the enemy. The same escapist delusion that possessed a few itinerant experimenters during the Raketenflugplatz days now infected the German military establishment. Albeit an engineering marvel, the V-2 stands as one of the twentieth century's most foolish and costly misapplications of resources in the name of advanced technology, contributing to the ultimate German catastrophe of 1945.[4]

Von Braun's official position in this program was technical director for development, which meant that he supervised the engineering work that ensured that missiles moved from drawing board to finished product. For such a complex device in mass production, this was a broad responsibility that overlapped at times with virtually every aspect of the project, including utilization of slave labor. He was of course not personally involved day-to-day with every step, but he was a key manager. Nominally a civilian, he reported as he had since 1933 to General (since June 1943) Walter Dornberger, aged forty-eight, who had been chief administrator of the army rocket program since 1936. As a military officer, Dornberger was also concerned with the operational deployment of missiles in the field, which was outside von Braun's box on the organizational charts, though they certainly talked about it and von Braun visited launch sites in the field.

In 1943, von Braun was still just thirty years old, a remarkably young age for his job by any standard. Indeed, he had been the leader of hundreds of white- and blue-collar workers in the rocket program since he was twenty-five. By all accounts he possessed sufficient technical acumen, but the authority to preside over so many lives in a military setting had to stem from more than this. It is fair to assume that an important source was the "Freiherr" in his birth name, which could still make heels click in German society and was a rarity in his workaday milieu much coveted by the Nazi Party, whose membership had always suffered—right up to the top—from social inferiority. He was also the physical epitome of "Aryan," to which years of racist propaganda had lent an aura of charismatic bio-logical perfection. Throughout his association with the army rocket pro-gram, even when he was a lowly "student," his iconic visage was included in group photographs of officers and dignitaries. Dornberger often brought him along to help sell the program up the bureaucratic ladder, much as Nebel had put him forward as a public face for the Raketenflugplatz when he was only a teenager. The Reichswehr was traditionally bejeweled with Prussian nobility, of course, though Dornberger was the son of a middle-class pharmacist.

In addition, von Braun was a high-ranking officer in the SS. His first association with the Schutzstaffeln had come while still a student in the fall of 1933, when he joined a Berlin *Reitersturm,* or SS horseback riding unit.[5] At a time when *Reichsführer-SS* Heinrich Himmler was recasting the organization away from its plebeian roots among Freikorps veterans, jobless intellectuals, and lower-middle-class Party stalwarts, he was espe-cially determined to woo the equestrian stronghold of blue-blooded Ger-man conservatism.[6] The "riding lessons," as von Braun referred to them many years later (which as a child of the landed Junker aristocracy he surely did not need), took place in a context of political indoctrination and the sartorial flummery of head-to-toe black uniforms tailored to drama-tize the formation of a fearsome Imperial Guard, but did not require entry into the ranked order of Himmler's cadre.

When he joined the Reitersturm, von Braun was already ensconced as an Army Ordnance-sponsored graduate student among a coterie of offi-cers and professors with tight connections to the Nazi Party. It is incon-ceivable that he would not have witnessed or at least heard of the *Blutkrieg* in Berlin's streets, or recent public events such as the national boycott called by Hitler in April 1933 of all Jewish businesses and professions, when most of the shops along Berlin's swank Kurfürstendamm were defaced with the word *Jude* in white paint and mobs stood outside the Ka De We department store chanting anti-Semitic slurs.[7] In May, fellow students from the University of Berlin helped burn more than 20,000 books at the Opernplatz on Unter den Linden while Goebbels broadcasted over a microphone that "the age of extreme Jewish intellectualism has now ended."[8] Under the "Aryan Clause" of a new civil service law, the faculty of every German university was being purged of members with undesirable politics or racial heritage. All Jewish academics and medical doctors were forcibly retired. As these and myriad other fulfillments of Hitler's promises rapidly unfolded in open view, seizing the world's alarmed attention, von Braun affiliated himself with the SS.[9] If he somehow felt pressured to join, he never said so in later life.

"The entire outfit never did participate in any activity whatever outside the riding school during my connection with it," he testified many years later, in toto, about the experience.[10] This statement was typical—and one of the most astonishing—of the tunnel-vision comments he made regarding his early life in Germany.

In April 1940, by which point the rocket research program in Berlin had blossomed into full-scale development work at Peenemünde on the remote Baltic Sea coast island of Usedom, von Braun was visited at his office by a local SS colonel, who was under personal orders from Himmler to secure von Braun's membership.[11] Production plans for a long-range missile were bogged down in wartime bureaucratic rivalries for increasingly scarce materials and labor.[12] High-level infighting during the procurement crisis was so treacherous that General Karl Becker, who had nursed and protected the rocket program since its earliest days in the 1920s and become head of Army Ordnance in 1938, killed himself, creating a top-level power vacuum at a precarious time.[13] This was also the season when hundreds of Polish workers arrived at Peenemünde as part of the exploitation of conquered populations in the East under Himmler's aegis.

Von Braun claimed many years later that he told the colonel that he was "so busy with my rocket work that I had no time to spare for any political activity," to which the SS officer curtly replied that membership "would cost me no time at all." Von Braun knew, nonetheless, that "the matter was of highly political significance for the relation between the SS and the Army." After stalling for time to consider the matter, he went immediately to Dornberger, who "informed me that the SS had for a long time been trying to get their 'fingers in the pie' of the rocket work" and that "if I wanted to continue our mutual work, I had no alternative but to join." It is unlikely that the SS's ambition was news to von Braun, since he had instantly gauged the real importance of the colonel's visit. On the first of May, he therefore became an SS *Untersturmführer* (second lieutenant) and received yearly promotions until reaching the rank of *Sturmbann-führer* (major) in June 1943.[14]

Whether or not von Braun's recruitment occurred exactly as he recounted, it is clear that Himmler had determined when and where to introduce his "fingers in the pie" and that this was a canny foray that neither von Braun nor Dornberger could refuse or even delay, if either had wished to do so. "Himmler very seldom expanded his responsibilities through protracted jurisdictional fighting," Albert Speer, the Reich Armaments Minister, remembered many years later. "Instead he would patiently lie in wait, and then suddenly spring into action when he saw his chance."[15] Though a so-called honorary appointment, which Himmler bestowed upon numerous professionals during his reign, the official insertion of von Braun into SS ranks—and, on important occasions, into SS uniform—at this juncture marked an existential shift for him within the competing satrapies of the Third Reich. The SS and the army were mutually distrustful hierarchies with their own chains of command in a martial universe where ideological imperatives devolving from the führer cult trumped conventional military authority.[16]

Speer wrote that he rejected an honorary SS rank in 1942 because it would have "put me into an unofficial vassalage to Himmler" and "sharply disrupted my confidential relations with the command posts of the army." Citing the example of an administrator of German government in occupied France and Belgium whom Himmler "snapped at" and told to pursue SS policies "because he was an honorary commander in the SS," Speer maintained that Himmler "exerted total pressure on his SS honorary commanders, whom he viewed as vassals pledged to him, even though he kept stressing that honorary rank in the SS involved no obligations."[17] Von Braun acknowledged the thrust of this when he recalled that Dornberger "hoped that our old cordial relation of confidence would avoid any future difficulties that could arise" from his new SS commission. A purely honorary, substanceless membership would not have triggered such concern.

Thus, by mid-1943, von Braun stood at the epicenter of a major munitions program with thousands of workers and ultimately worth billions of marks, being rapidly drawn into the polycratic leadership intrigues and incoherent decision-making at the highest level that characterized the remainder of the war. A futuristic weapon never before tried in warfare (Dornberger's starting point for envisioning the military value of a "big rocket" was the Great War's so-called Paris Gun, a long-range cannon that fired shells 25 miles high to reach targets 75 miles away, but proved to be of negligible worth), the V-2 required astute bureaucratic salesmanship, amounting to exaggeration of its potential to help win the war.[18] More Nebelesque Raketenflugplatz bunkum, now dressed in eminent Reichswehr *feldgrau*.

The man who perhaps had been most resistant to this brand of persuasion was Hitler himself. He had first seen Dornberger's operation during a tour of the Kummersdorf proving grounds in September 1933, with a group of officers and Party luminaries that included Air Minister Hermann Göring and Interior Minister Wilhelm Frick. A second visit in early 1934 brought along Deputy Führer Rudolph Hess and his then-personal-secretary Martin Bormann, plus various Brownshirt leaders at a time when tensions were rife between the Reichswehr and the SA (storm troopers) over which organization would form the future army of the Third Reich.[19] Von Braun appeared in group photographs on both occasions, but there is no record of any direct personal encounter between the führer and the handsome young baron.[20] On his way to fulfilling his promises of rearmament and abdication of the Treaty of Versailles, but still years away from the realities of war, Hitler was perusing an important weapons development center and opening coffers that had previously been limited. "Our need," Dornberger wrote many years later about this time, "was for higher authority to give our work due recognition and to provide us with money—a great deal of money—and with the staff for carrying on."[21] There was now only one such "higher authority" in Germany.

In this regard, von Braun no longer held the close family link to national government that had so recently helped open doors for him. Franz von Papen had lost the chancellorship in December 1932 to Kurt von Schlei-

1. Rudolf Nebel (*far left in white coat*) stages a publicity photo to mark the July 23, 1930, test of his rocket-enthusiast group's *Kegeldüse* engine in Berlin. Spaceflight author Hermann Oberth stands to the right of a large model, probably a prop from the movie *Frau im Mond*. Klaus Riedel, the group's designer, holds a conventional signal rocket for dramatic effect. Behind him in knee-pants stands eighteen-year-old Wernher von Braun.

2. A 1940 German cutaway drawing of the A-4 rocket, later dubbed the V-2 (for *Vergeltungswaffe*, vengeance weapon) by Nazi propaganda minister Josef Goebbels.

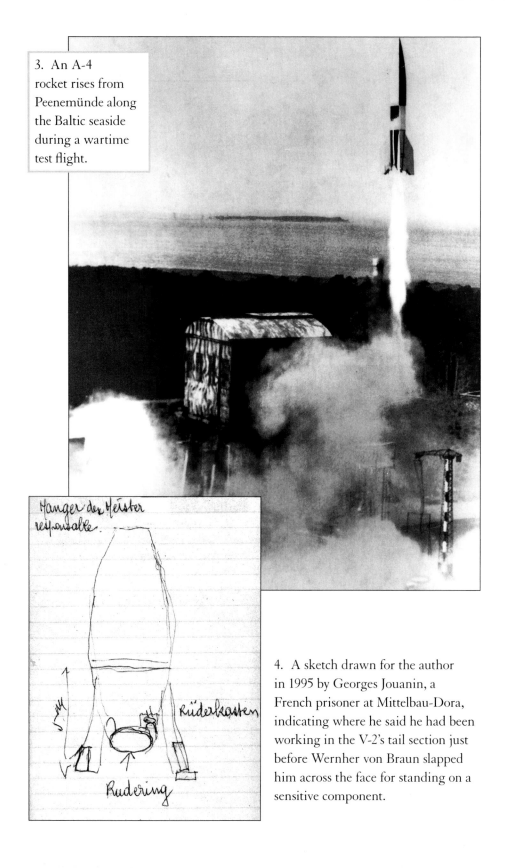

3. An A-4 rocket rises from Peenemünde along the Baltic seaside during a wartime test flight.

4. A sketch drawn for the author in 1995 by Georges Jouanin, a French prisoner at Mittelbau-Dora, indicating where he said he had been working in the V-2's tail section just before Wernher von Braun slapped him across the face for standing on a sensitive component.

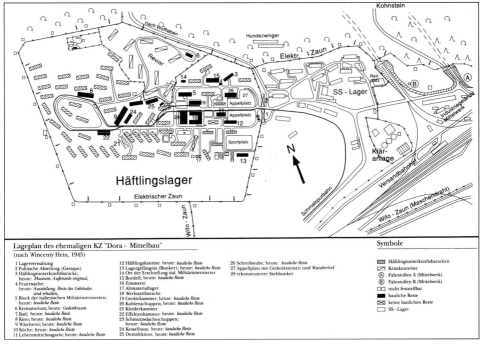

5. A German map of the Mittelbau-Dora slave labor camp and underground rocket factory. The prisoner camp (*Häftlingslager*) is on the left, surrounded by an electrified fence. Outside the fence to the right are SS guard barracks and factory tunnel entrances (*at positions* A *and* B). The crematorium (*at position* 6) is one of the camp's few surviving structures today.

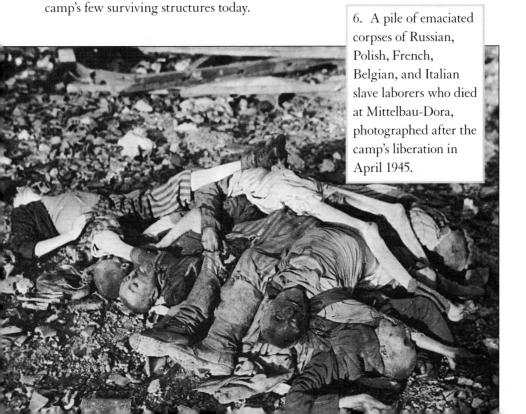

6. A pile of emaciated corpses of Russian, Polish, French, Belgian, and Italian slave laborers who died at Mittelbau-Dora, photographed after the camp's liberation in April 1945.

7. American soldiers pose with German "rocket scientists" after their carefully choreographed surrender in the Bavarian Alps on May 2, 1945. Wernher von Braun stands at center with arm in cast; Walter Dornberger is next to him holding a cigarette. Pfc. Frederick P. Schneikert, who first encountered them as an interpreter with the U.S. Army's 44th Infantry Division, is the shorter G.I. wearing a helmet (*left*).

8. Homeless German women with a child, living next to V-2 rocket engines in one of the abandoned factory tunnels at Mittelbau-Dora after the camp's liberation.

9. A liberated French slave worker, armed but still wearing his striped prisoner suit, shows rows of camouflaged V-2 fuel tanks to an American soldier at Mittelbau-Dora.

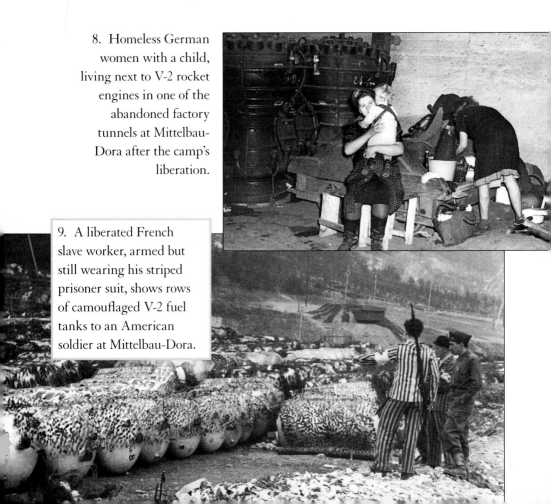

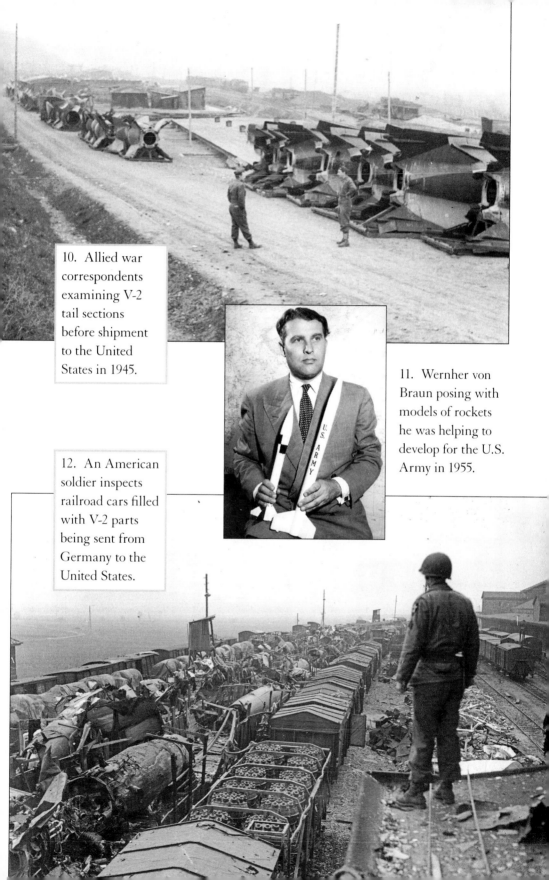

10. Allied war correspondents examining V-2 tail sections before shipment to the United States in 1945.

11. Wernher von Braun posing with models of rockets he was helping to develop for the U.S. Army in 1955.

12. An American soldier inspects railroad cars filled with V-2 parts being sent from Germany to the United States.

13. Wernher von Braun poses in 1959 with his Mercedes 220SE in front of the Army Ballistic Missile Agency's Redstone Building in Huntsville, Alabama.

14. Wernher von Braun rides with his daughters Magrit and Iris in Huntsville's "Moon Day" parade on March 4, 1959, celebrating the launch of the *Pioneer IV* satellite.

15. Wernher von Braun shakes hands with President John F. Kennedy at the Marshall Space Flight Center on September 12, 1962. Alabama Congressman Robert Jones is between them, with Vice President Lyndon Johnson partially visible behind Kennedy's shoulder. The president's older brother, Joseph Jr., was killed in August 1944 during a bomber mission against a V-weapon site in France.

16. Saturn V rocket on its launch pad being prepared for the *Apollo 11* moon mission in May 1969 at Cape Canaveral, Florida, showing striking resemblance to Peenemünde.

17. Huntsville officials carry Wernher von Braun on their shoulders, celebrating the splashdown of *Apollo 11* on July 24, 1969.

18. "Wernher von Braun Day," February 24, 1970, at the Marshall Space Flight Center, Huntsville. Alabama Senator John Sparkman stands between Alabama Governor Albert Brever (*left*) and Wernher von Braun.

cher, a general turned defense minister who had parlayed with Hitler and
Röhm in 1931 about joining the Weimar government.[22] The Nazi Party's
electoral strength was actually slipping, but the threat of civil war and the
pressures of a comatose economy kept it at the forefront of national politics. Short-sighted intrigues among competing supporters of Schleicher,
Papen, and Hindenburg resulted in Hitler's being handed the chancellorship at the end of January 1933 in a horrendously misguided attempt to
contain him within a circle of traditional conservatives. Magnus von
Braun's job as minister of agriculture passed to far-right nationalist Alfred
Hugenburg, who lasted only until June 1933, when he was replaced by
Hitler's longtime agricultural activist, Walter Darré.

The elder von Braun did not depart out of anti-Nazi principle. To the
contrary, he wrote in his memoirs that he would have preferred to continue under Hitler, because he saw no other course ("ich hätte keinen
anderen Ausweg gewust").[23] Though permanently retired from the corridors of power, he and his wife Emmy did not absent themselves from the
glittery social events they had been attending since *Kaiserreich* days that
now spotlighted Nazi leaders. One of his closest friends, Franz Gürtner,
carried over into the Third Reich as justice minister and visited the von
Brauns in Berlin and at their country estate for years while he counte-
nanced Hitler's erosion of the law.[24]

But Wernher no longer needed his father. He completed his studies in
June 1934, having progressed from Leitz schoolboy to Doctor of Philosophy in four years.[25] He was just twenty-one years old. He had spent
the last year and a half at Kummersdorf in "one-half of a concrete pit
with sliding roof" situated between two artillery ranges, helping devise
small liquid-fueled rocket motors with what he called a "staff" of one
mechanic.[26] Work was initially based on what they could remember
from the Raketenflugplatz, using alcohol and liquid oxygen as propellants, which was the only way he knew how to start. "To the amazement
of the authorities," the first test was a success, he recalled many years
later. But the most proximate of those authorities, Walter Dornberger,
had a polar opposite memory, describing in vivid detail how the test
stand was wrecked by a violent explosion. Dornberger also mentioned

someone other than von Braun as filling the role of "test engineer and designer." That someone was Walter Riedel, an experienced engineer hired from the Heylandt company, where Max Valier had continued tinkering with rocket cars until his death.[27] Riedel "seemed to me to provide the right counterpoise to the rather temperamental—and at that time self-taught—technician, von Braun," Dornberger remembered, adding that the seasoned Riedel repeatedly managed to guide the bubbling stream of von Braun's ideas into steadier channels."[28] Dornberger obviously felt that von Braun was still wet behind the ears and that the road ahead required at least one man who knew what he was doing.

The memoirs of von Braun and Dornberger agree only that the project was plagued by constant trouble—not unexpected in such development work, of course. In six months they built a potentially flyable rocket designated A-1 (for *Aggregate*, or unit), about 4.5 feet long and 1 foot in diameter, but "within half a second of firing it burst into fragments."[29] A second A-1 with a more dependable motor was set aside when they realized that the flywheel mounted in its nose would not help to keep it from tipping off-course during flight.

Von Braun's doctoral dissertation title page read *Konstructive, theoretische und experimentelle Beiträge/zu dem Problem der/Flüssigkeitsrakete.*[30] It has often been rendered into English as the awkward-sounding "Design, Theoretical and Experimental Contributions to the Problem of the Liquid-Fueled Rocket," but the word *Konstructive* does not translate cleanly as "design" in the sense of original conception. Von Braun probably did not design much hardware; that was Riedel's job according to Dornberger. In the thesis text, von Braun focused on combustion theory and measurements, more in keeping with his lack of practical skills. In the era of Einstein and Heisenberg, there was of course no advanced theoretical physics, though this would have meant nothing to his advisers.[31] In any case, his Friedrich-Wilhelm diploma carried the deliberately obtuse title *Über Brennversuche*, "About Combustion Experiments," because the army wished to keep it all secret. In April, Goebbels's propaganda ministry had outlawed public mention of any military or technical aspects of rocketry.

In addition to the classification of his doctoral work, von Braun personally encountered the blanket of secrecy being dropped around German rocketry when the gestapo interrogated him a month later about his relationship with Rudolph Nebel.[32] The irrepressible majordomo of the Raketenflugplatz had been warned by Karl Becker's office in the latter part of 1933 to stop bantering in the press and elsewhere ("unerwünschte Propaganda und Presseveröffentlichungen") about rockets as weapons.[33] During Hitler's "Night of the Long Knives" purge of the SA between June 30 and July 2, 1934—when scores of Brownshirt leaders and prominent political figures (including recent chancellor Kurt von Schleicher) were executed by SS gangs—Nebel had been jailed at gestapo headquarters in Berlin, where he evidently mentioned von Braun as his inside man at Army Ordnance.[34] Von Braun countered that he had conveyed to Nebel via several meetings with Klaus Riedel in 1933 that Ordnance's door was shut and that, for Nebel's own sake, he should stop the publicity ("Nebel im eigenen Interesse, sowie im Interesse der Sache Veröffentlichungen von technischen Details und von der militärischen Verwendbarkeit lassen solle").[35] The army stood behind von Braun, and Nebel managed to carry on for years with his quixotic obsession. In retrospect, the episode served to illustrate the disjointed political and military fiefdoms that were already at odds within the Third Reich.

Not until December 1934 were products of the effort at Kummersdorf successfully launched. Christened "Max" and "Moritz" after the mischievous duo in Wilhelm Busch's 1865 children's classic, two "A-2" sounding rockets rose about a mile high from sandy Borkum, a 12-square-mile island situated on the western end of the same North Sea coastal chain as von Braun's alma mater at Spiekeroog. This was what Hermann Oberth had been hired to do in 1929 for *Frau im Mond* cinema publicity, but it was undeniable progress.[36] Their motors were essentially the same as the A-1's, but the flywheel-cum-gyroscope had been moved from the nose to the middle, between the oxygen and alcohol tanks, to create a more stable center of gravity.[37] Their fabrication had benefited from the addition of another engineer at Kummersdorf with industrial experience, Arthur Rudolph, who took charge of the workshops.[38] Dornberger was beginning to assemble a self-sufficient team under government auspices, independent of commercial contractors, shrouded in secrecy, that could go beyond the traditional testing function of military arsenals like Kummersdorf into actual development of weapons—a system that would come to be known as *Alles unter einem Dach*, "all under one roof."[39] The Borkum performance underlined, however, that the military rocket program was physically outgrowing Kummersdorf. "We had big desires," Dornberger remembered.[40]

Max and Moritz were not weapons—even if they had been somehow customized to carry small quantities of the era's high explosives or poison gas, they would have offered the army nothing that artillery shells could not already do better. They made sense only as stepping stones to truly long-range guided missiles that could travel for hundreds of miles with heavily damaging payloads. If there had ever been any ambiguity or obliviousness in von Braun's young mind about what exactly he was working toward, it necessarily ended now as he participated in what Dornberger called "the old dodge" of multimedia shows—lectures illustrated with films, color drawings, and diagrams—of "our wares in front of the prominent people who sit on the money bags."[41] If an important general could be persuaded to witness a live firing of the thunderous rocket motors, the effect was usually to enrapture.[42]

In January 1935 the Nazi government enjoyed its first foreign policy coup when a plebiscite in the Saar region—a coal-rich borderland between France and Germany that had been occupied jointly by the British and French under the Treaty of Versailles—resulted in an overwhelming majority of voters favoring a return to Germany.[43] At the beginning of March the British government issued a report to Parliament calling for more weapons production in response to known German rearmament that was illegal under Versailles restrictions. When Hitler answered by announcing that he would inaugurate a German air force, which already existed behind flimsy secrecy, there was scant international protest. He then took the bolder step of reviving compulsory military service and proclaiming the "restoration of German sovereignty in military affairs," thereby splintering the treaty once and for all.[44] As usual, he syncopated such belligerent moves with dramatic peace speeches, denying all plans of war and offering to sign nonaggression pacts.

In this blustery atmosphere the German rocket program's budget began to expand to the proportions Dornberger wanted. The Luftwaffe's sudden prominence under the blue-sky ambitions of Hermann Göring soon added even more momentum. "In January 1935 we were visited by Major von Richthofen [Wolfram Freiherr von Richthofen, 1895–1945], a cousin of the great ace of World War I," von Braun recalled years later.[45] "Hitler's power was rising and the Luftwaffe became the recipient of a degree of generosity not extended to other branches of the Wehrmacht. Luftwaffe officers up to the rank of general were young, enterprising, and receptive and didn't suffer from the hidebound mentalities and masses of red tape which handicapped the Army and Navy."[46] During Dornberger's absence on another assignment for the artillery corps, von Richthofen barged through service protocol and engaged von Braun's interest in developing rocket-propelled fighter aircraft, which got underway in the summer of 1935. "Von Richthofen evinced not the slightest embarrassment at asking an artillery experimental station to develop an aircraft power plant," von Braun recalled. "No obstacles were placed in the way of our accepting the job."[47] It was the first time, but not the last, that von Braun would experience the ability of men who somehow carried the führer's special blessing to circumvent traditional lines of authority. Letters that he wrote from his *Heereswaffenamt* office in Charlottenburg now carried his signature under the closing "Heil Hitler!"[48]

Von Richthofen's project could not be accommodated at Kummersdorf. "Nothing daunted, the doughty Major promptly offered us five million marks for building more ample facilities at some other location. His offer constituted an unprecedented breach of military etiquette between branches of the Wehrmacht."[49] But Arthur Rudolph held a firm memory of a different genesis for the Luftwaffe's monumental offer, one that featured a "free-wheeling" von Braun operating on his own without Dornberger around. After informal conversations about Kummersdorf's inadequacies, according to Rudolph, von Braun told him that his mother had suggested Peenemünde as a perfect place for a much larger facility. "What I definitely know is that von Braun said, 'Let's go and see the *Reichsluftfahrtministerium*, somebody in there,' and we got there and we were received by a *Ministerialrat* or *Ministerialdirektor*," Rudolph recalled, identifying the official as Adolph Bäumker, head of aeronautics research at the Air Ministry. "Von Braun in his usual manner made an excellent presentation to him. And before he even finished, Bäumker said, 'Von Braun, I give you five million of Luftwaffe money [about $12.25 million] so you can start the ball rolling.' "[50]

Whether it was Freiherr von Richthofen or Freiherr von Braun who breached military etiquette, or perhaps both, cannot be known.[51] In retrospect it is apparent that without Dornberger, the rocket program was drifting away from its conceptual roots in long-range artillery toward an application for aircraft, which would never have yielded a ballistic missile. The young von Braun's love of flying—he took lessons again in 1934 and 1935—and infatuation with Luftwaffe élan seemed to be as strong as his childhood dream of space travel. By his own admission, von Braun wound up in the office of a "wrathfully indignant" General Karl Becker. On the spot, according to von Braun's memory, Becker said that he would appropriate 6 million marks [about $14.7 million] in addition to von Richthofen's five million. Von Braun never mentioned a free-lance visit to Bäumker's office or Becker's reaction to it.

"In this manner our modest effort whose yearly budget had never exceeded 80,000 marks, emerged into what the Americans call the 'big time,' " von Braun concluded. "Thenceforth million after million flowed as we needed it."[52]

Dornberger reappeared almost immediately, in March 1936, as chief of the Ordnance rocket program, indicating that Becker wanted him back to preside over the burgeoning enterprise. His rank rose to major and the program to a higher rung on the bureaucratic ladder that enabled it to request funds on its own as one of a dozen independent sections within the army's R&D establishment. At the beginning of the month, Hitler scored a profoundly significant triumph by marching German troops into the Rhineland—against the strong advice of his military and diplomatic leadership—with no subsequent French resistance. An exhilarated nation now supported his dictatorship overwhelmingly (as the gestapo continued to crush any opposition and the conservatives who had brought him into power stood aside). The commander-in-chief of those forces, General Werner von Fritsch, reacted to one of Dornberger's dramatic lecture shows and live rocket demonstrations by asking "bluntly and dispassionately" the "all-important question, 'How much do you want?' "[53]

Another meeting followed in April with General Alfred Kesselring, an artillery veteran who was now the Luftwaffe's construction czar, which included Becker, von Richthofen, and von Braun. Dornberger had already outlined with von Braun and Walter Riedel the rough specifications for a long-range rocket to be developed at Peenemünde: 1-ton payload, 25- to 30-ton thrust engine, body dimensions that would allow it to be transported by road or through any railroad tunnel in Europe. Considering that the only success they had achieved so far was with the A-2's 300-kilogram thrust engine, this represented a fantastic order-of-magnitude leap. They also wanted to construct factories and housing at Peenemünde "on the grand scale, and beautifully," not like the niggardly architecture of old army buildings. "Kesselring could not help smiling at our enthusiastic and even dramatic picture of the future," Dornberger remembered, but finally offered to pay half the cost. Demonstrating either Kesselring's excitement or perhaps fear of real estate corruption, an Air Ministry official was dispatched that same day "in a high-powered car" to buy the required land on Usedom from the nearby town of Wolgast for 750,000 marks. "Here was action indeed!" even Dornberger marveled.[54]

The rocket program and the Third Reich were entering the headiest years of their brief existence.

7

SUPREME ZEAL

U NLIKE THE THURINGIAN environs of Mittelbau-Dora, which even
on infrequent sunny springtime days possess a grim air, Peenemünde
retains some of the clarity of the Baltic seaside. Until the 1930s, this was
marshy backcountry with a plenitude of wild ducks, geese, and Pomera-
nian deer that attracted few people other than hunters. A fishing village of
thatched cottages and some 450 inhabitants at the mouth of the Peene
River (hence *Peenemünde*) was the only sign of civilization, though fashionable resorts like Zinnowitz and Ahlbeck—"strung along the coast like
a necklace of pearls," as Walter Dornberger fondly remembered—were
not far away. Piney woods, rolling surf, and white sand beaches would
remind an American of Cape Cod or, if it were all somehow transferred
to Florida, Cape Canaveral.[1]

In 1936, the area was a pristine canvas upon which would rapidly be
painted a National Socialist dreamscape of state-funded research and
development. Since the latter half of the nineteenth century, Germany
had pioneered fertile relationships among academia, industry, and government that resulted in world leadership in science and engineering.
During the Great War, the military had profited from this system's innovations, such as the Haber-Bosch process for synthesizing ammonia, which
enabled mass-production of explosives after the Allies blocked nitrate

shipments from Chile. Peenemünde's creators envisioned it as emblematic of the new Germany, a Technik und Kultur showcase whose architecture would reflect the boundless future, where the best of everything would be on hand in a beautiful setting to produce bold leaps of progress, at least for the military.

Indeed, there was nothing like it anywhere else in the world until the Manhattan Project became the new model for such endeavors. By then, only the shards of "mein schönes Peenemünde" would be left in the detritus of the Third Reich.[2] That is what a visitor finds today around the broken walls of a massive liquid-oxygen production plant, or deep in the pine groves where mangled rebar and flooded concrete foundations of destroyed launch pads rest beneath thick undergrowth, booming surf in the background, reminiscent of Mayan ruins. It is the scale of the place that still impresses. Though the rockets were small compared to what lay in the future, their inventors were the first to comprehend the spatial dimensions that manipulating so much chemical energy would require.

Peenemünde was never a public edifice like Fritz Haber's renowned *Kaiser-Wilhelm-Institute für physikalische Chemie und Elektrochemie* (Kaiser Wilhelm Institute for Physical Chemistry and Electrochemistry) in Berlin, however.[3] It was devoted purely to weapons development and had no civilian function. One of the primary reasons for being located in a remote wilderness was to maintain secrecy about what went on there. In the longer run, its separateness would reinforce the elitist notion that science can somehow be conducted without connection to the society around it, that the scientists and engineers themselves can be untainted by politics. "The Army wisely and effectively prevented political interference at Peenemünde by pleading the need for secrecy," von Braun insisted many years later, though he himself wore a Nazi Party swastika button on his lapel from the beginning.[4]

The men who lived to tell about it after 1945—von Braun, Dornberger, and others—insisted self-defensively that space travel was their ultimate dream (and why not, presented in this utopia with a 4.5-ton rocket, 46 feet tall?). But there was never a day when their work was not focused on building weapons for the Hitler regime, or when they did not subscribe to policies that were brutally suppressing the free practice of science in Germany. The crown jewel of German R&D from 1937 to 1945 was a Nazi armaments lab, not a center for open exploration.

Of course, besides what Hitler would need to conquer and racially cleanse the world for German expansion, he had no vision for the future beyond grandiose monuments. In fact, what Peenemünde finally produced—a finicky device that required Croesusian expenditures and constant obsessive attention to technical minutia—achieved nothing in strategic terms other than an expansion of the Great War's artillery capabilities. Meanwhile, the United States and the Soviet Union mass-produced dependable aircraft and conventional munitions that eventually turned the tide against Hitler. Historians have parsed Germany's narrow-minded production of unreliable rockets from many angles ever since, but it is still reasonable to wonder simply whether the rocket's magical control over prodigious amounts of energy not only astonished the generals into buying it but also seduced the young technologists into perfecting it. Zeal was supreme.

WHILE CLEARING AND construction got underway at Peenemünde, with the first crews wielding axes and two-man saws, Wernher von Braun satisfied his military service requirement at a Luftwaffe flying school near Berlin between the first of May and the end of July 1936. Given that Göring's new air force was footing half the bill for the *Heereversuchsanstalt* (army experimental laboratory), this was both politically smart and pleasurable for the twenty-four-year-old slated by Army Ordnance to front its engineering workforce there. The German public's ongoing fascination with aviation had always been stronger than any fad for rocketry during Max Valier's brief heyday. Von Braun had already expressed his own enthusiasm by learning how to pilot gliders and propeller-driven airplanes.[5] If the rocket program had not been rooted in established army artillery circles thanks to Becker and Dornberger, the natural place for it in Germany's vernal *Aufschwung* (revival) would have been the future-oriented and financially flush Luftwaffe. The old artillerists made sure this did not happen, and putting a youthful face—a handsome Aryan one, moreover—on an army program for good measure was perhaps another reason why von Braun's early career seemed to defy gravity. It also helped ensure that he did not defect to the air force, which by all indications would have welcomed him and not cared about protocol.

Reflecting the funding arrangement, the military enclave on Usedom split into two parts known as Peenemünde-East and Peenemünde-West, for the army and Luftwaffe, respectively. Peenemünde-East ran southeast from the island's northern end along the Baltic beaches to the resort village of Karlshagen, a distance of about 8 kilometers. Testing long-range rockets could best be done by launching them from the shoreline out across the open sea, secretly (for awhile, at least) and safely. Peenemünde-West, which consisted mostly of an airfield and hangars, occupied the northwestern flats that faced the Greifswalder Bodden, a shallow bay between Usedom and the much larger island of Rügen. Administration buildings, laboratories, and housing units—a worker's *Siedlung* (settlement or housing estate) was raised near Karlshagen, connected to the base by rail—dotted the open spaces. "The architecture was attractive, combining a resemblance to the older municipal buildings of the northern provinces of Germany with a touch of the twentieth-century Bauhaus school," remembered Peter Wegener, an aerodynamicist who arrived in 1943 and found "a bucolic atmosphere" after service on the Russian front. "In contrast to most other architecture during Hitler's time, as exemplified by Ludwig Troost and Albert Speer, the modern style had not been completely abandoned here."[6] Finding even a touch of Bauhaus was perhaps wishful retrospective thinking on Wegener's part. Peenemünde was meant to be a model of National Socialist esthetics. The official entrance to the base was marked by the stone *Berliner Tor*, a triumphal arch in heavy New Order neoclassical style replete with swastika looming overhead.[7] It was nicknamed the *Brandenburger Tor*.

Seventy years later, von Braun would still be described as the heart and soul of Peenemünde, its star and essential figure.[8] The memoirs of men who worked with him there and then came to the United States under his wing are so exquisitely unanimous in this regard—except for Dornberger's, which sometimes mentions von Braun's shortcomings and contains a photograph of him in his SS uniform (albeit obscured)—as to give pause to an impartial observer. Almost every actual hands-on task of design and production was performed by someone else, with von Braun weighing in as a coordinator and facilitator—not necessarily afraid to get his own hands dirty, but supervising. Many of his subordinates spoke glowingly in retrospect about his perfect manners and ability to chair a meeting, while noting his forceful impatience, sometimes meanness, with anyone he felt came unprepared or attempted a "snow job." This was a powerful role in a complex organization, what today would be called systems engineering, but von Braun did nothing that Dornberger did not approve or arrange as his superior. He was too young to have acquired venerable experience, which is why Dornberger hired seasoned men from the start to carry the project forward on the shop floor. When Peter Wegener marveled how von Braun "could separate important from peripheral items, distinguish what had to come first, make clear decisions, and inspire people to work," it sounded strangely at odds with von Braun's chronological age (31), even in 1943.[9]

Long after the war, Dornberger wrote a letter praising von Braun as a "systems engineer who knew how things have to and will fit together." He made clear that the young man was not a wellspring of original work, however: "I know that [von Braun] during the years in Germany never wrote alone by himself an outstanding treatise about a scientific subject, which in its analysis and newness, with its conclusions and results shook the scientific community. I know also that he personally never designed down to the last minute detail any technical device, which in its perfection has been unique and an outstanding engineering break-through. I also know that his personal managerial talent was not without faults and mistakes. I know that his undisputable ability as a leader was not always perfect. However, his personal contributions in all fields of modern rocketry [were] innumerable, decisive, guiding, leading."[10]

If such appraisals are accepted at face value, then von Braun was one of the most precocious wunderkinder in the history of technology. Indeed, this was his public celebrity reputation for the rest of his life.[11] But if they are taken instead as heroic hyperbole, then their psychological dimension must be limned. Why did some men adore him? It is fair to posit that he served as an idol, one who could be adulated and yet also absorb guilt much as the führer channeled away personal responsibility for an entire nation. If von Braun was the essential figure, then everyone else was not.[12] It is also plain that because the Peenemünders who later lived in America were dependent on him, they feared him for what he was in a unique position to assert about their professional value as "intellectual reparations," as well as their past conduct.[13] In any case, if any of these men held contrary opinions about von Braun, they kept silent.

Thousands of others who did not make it to America were never heard from at all, of course.

THE SUMMER OF 1936 marked Germany's prideful revitalization after the humiliations of Versailles, the political knots of Weimar, and the economic misery of the Great Depression—an upswing epitomized by the pageantry of "Hitler's Olympics" in August. Now the Nazi Party's thuggish Horst-Wessel anthem was performed for an international crowd of 110,000 people in the world's largest stadium by a 3000-voice choir conducted by Richard Strauss (though in January 1936, in anticipation of a flood of foreign visitors, extreme anti-Semitic propaganda was prohibited for the duration of the games). Most Germans—except of course Jews (about three-quarters of 1 percent of the total population of 67 million), leftists (nearly extinct), and others deemed undesirable—admired the führer at this juncture for bringing about welcome changes in every sphere of life from the workaday to the global, even though relatively few matched his fanatical obsessions.[14] The only cloud in the sky was a conflict between finding sufficient resources for armaments and consumer goods. Since Göring was taking control over economic policy, however, this posed no problem for Peenemünde. Hitler's preoccupation with "Jewish Bolshevism" as a threat to German mastery in Europe led him to support anti-Republican rebels in Spain under General Francisco Franco, adding impetus to militarizing the economy for a war that he thought was inevitable (and closer than his generals believed). By the end of the year, with the formation of an "Axis" between Berlin and Rome and a pact with

Japan, the menace of expansionist power was becoming palpable on a wide scale.[15] Germany seemed hell-bent on rearming.

After finishing his Luftwaffe stint, von Braun returned during the Olympics to Berlin and Kummersdorf, where work continued on the next stepping stone toward a long-range missile, designated A-3. Still "purely experimental" and not equipped to carry any payload except instruments, according to Dornberger, it was nonetheless a step up in size and thrust at 22 feet long and 1500 kilograms, respectively—a big, heavy device for its day.[16] Once again, Walter Riedel designed the all-important engine, with Arthur Rudolph in charge of construction. "Von Braun listened to the two of us, and if we had any disagreements, he would resolve them," remembered Rudolph, underscoring von Braun's role as a manager.[17] Whether the two older and more experienced men ever colluded to steer such meetings in the directions they needed cannot be known. Von Braun was, after all, not much beyond the "Sonny Boy" stage of life.

Thus, though lavish resources were already committed in 1936 to construct facilities at Peenemünde for building ballistic missiles of military value, the Kummersdorf group held in hand only another sounding rocket like "Max" and "Moritz," albeit larger. They were perfectly aware that serviceable weapons were still far beyond state of the art. As they continued "pestering" higher army authorities for money, Dornbeger recalled, they were told that they would only be supported for rockets that could carry large quantities of explosives over great distances "with a good prospect of hitting the target"—that is, for reliably destructive weapons, a rational demand from military leaders. As they sketched out the parameters for such rockets, Dornberger added, they became "a bit uneasy" about being "a little too ambitious." In their "youthful zeal," they promised everything and ignored how much time it might take—an expression of blue-sky optimism in keeping with the heady atmosphere of Aufschwung.[18]

The A-3 incorporated new, laboriously tested ideas of how to combine and burn the rocket's propellants more reliably and efficiently, but was not basically different from the A-2. Its most vexing improvement was a gyroscopic control system to keep the missile pointed straight up as it flew aloft, the body being now too long and heavy to control with the brute force of a simple flywheel. Since "few of us had any knowledge of gyroscopy," von Braun recalled, the Kummersdorf group had for several years tapped the expertise of the Kreiselgeräte gyroscope equipment company near Berlin, a covert *Reichsmarine* firm that specialized in naval navigation, torpedo, and gunnery gear.[19] One of Kreiselgeräte's directors, Austrian navy veteran Johannes Maria Boykow, provided preliminary designs for the A-3 "black box"—a collection of gyroscopes working on three axes to register changes in pitch, roll, and yaw, plus accelerometers to sense sideways motion caused by wind gusts. Dornberger welcomed Boykow as a "clear-thinking scientist and practical man," terms he never quite applied to von Braun, whose imagination instead impressed him.[20] Electrical signals from the control apparatus rotated heat-resistant molybdenum steering vanes mounted directly in the rocket's exhaust flame. It was a promising mechanism, but Kreiselgeräte's forte was heavy shipboard components, not rocketry's novel realm, and every step needed strenuous testing.

The external geometry of the rocket's body had not yet been refined much beyond contours familiar to science-fiction fans and audiences of *Frau im Mond*. It was based, more or less as an educated guess, on the shape of an infantry bullet.[21] The A-3 could reach velocities greater than the speed of sound where air becomes compressed into shock waves— but there was no practical information about controlled flight other than the ballistics of artillery projectiles. Tail fins were necessary for arrow-stability, but their proper shape to help keep the rocket headed straight into the airstream through a range of altitudes and velocities was unknown. A supersonic wind tunnel was therefore needed for testing the aerodynamics of various forms (as well as how they heated up due to air friction), but the only one available to the Kummersdorf group was a miniscule 10-by-10-cm in cross-section, operated for the Luftwaffe by the Technical University of Aachen. At the end of September 1936, after a young professor from the faculty there, Rudolf Hermann, supplied worrisome results from testing tiny models of the A-3, Dornberger decided to seek a new supersonic tunnel, "cost what it might," for Peenemünde. Von Braun had been drumming for this, Dornberger recalled, but the cost scared him. The estimate was 300,000 marks, though Dornberger thought the final figure would reach a million, "especially with von Braun about."[22]

Dornberger took the wish up the chain of command to Becker, who "looked grave" when told the price. Becker eventually went along, as he always did, but stipulated that at least one other of the twelve sections within the army weapons R&D bureaucracy must also want the expensive tunnel. When Dornberger queried these departments, however, every one—including the Ballistics and Munitions office that had first harbored the rocket program—refused to endorse the purchase, indicating perhaps the sharp competition for funds or smoldering doubts within the regular army about the practicality of futuristic rockets. He finally won the vote of the antiaircraft artillery section, whose chief he knew personally. Rudolf Hermann, whom Peter Wegener would remember as "a serious disciple of Hitler," signed on to direct the new wind tunnel, which then began to take shape in Usedom's pine forest.[23]

Hermann was just one of an expanding cadre of experts hired from outside Kummersdorf's original small circle as Peenemünde gathered momentum. Not all of them possessed credentials of Hermann's quality—von Braun recalled that he even brought four men from the defunct Raketenflugplatz "back to their beloved work." They had been herded since the summer of 1934 into the aircraft instrument division of Siemens, the electrical industry goliath, with which Kummersdorf "maintained cordial relations," as von Braun put it. That is, the army placed them in jobs there as part of the concerted effort to remove rocketry from public view after Hitler came to power, and now was taking them back off the company's payroll. Rudolf Nebel, on the other hand, never got an invitation to work at Peenemünde. "It may be doubted whether his genius at salesmanship would have found a place in such an organization," von Braun quipped about his former mentor's braggadocio.[24]

Perhaps the most valuable researcher acquired at this time was Walter Thiel, a chemical engineering PhD who was already working for the army. His specialty was combustion, ostensibly the same field as von Braun's doctoral thesis, but his talent for turning theory into practice quickly overtook what von Braun had accomplished with the A-1 and A-2 engines. Two years older than von Braun, with similar fair hair and strong chin, though lacking aristocratic charisma, Thiel won Dornberger's highest approval as "extremely hard-working, conscientious, and systematic." Despite a "superior attitude" that made him a difficult colleague, Thiel became the dynamo behind myriad improvements to the A-3 engine and was given the task of designing the far more powerful motor for long-range rockets to be produced at Peenemünde. "Thiel's investigations showed that it required hundreds of test runs to tune a rocket motor to maximum performance," von Braun stated many years later, as though Thiel did little more than confirm the exasperating trial-and-error experience of Walter Riedel's and von Braun's earlier work. Von Braun found the prospect of such toil for the big rocket to be "frightening." With the arrival of disciplined technologists like Hermann and Thiel, von Braun's peripatetic imagination and Dornberger's shrewd ambition were being brought down to earth in a state of anxiety over promises made for the future.[25]

On January 30, 1937, to celebrate four years as chancellor, Hitler gave a three-hour speech before the supine Reichstag declaring that Germany would work in partnership with other nations to overcome Europe's problems.[26] That same month, he awarded Albert Speer the task of transforming Berlin into mega-monumental "Germania." In April—the month when the Luftwaffe's Condor Legion, under the command of Wolfram von Richthofen, bombed the defenseless Basque city of Guernica—Peenemünde began to receive its new occupants, who "seemed almost lost in the tremendous plant," von Braun remembered. If the thought crossed their minds at all, they might have wondered what kind of military project would require so much space, even more than the Hamburg shipyards that built great battleships. But Hitler's glory was at its zenith and there were few doubters in active circulation.

Though Peenemünde was von Braun's dream, he did not find its premises fit for living, instead choosing to settle several kilometers down the coast in stylish Zinnowitz.[27] He would remain there until 1939, despite the availability of comfortable housing on base. For a young bachelor with movie star looks, the attractions of the beach resort in season must have been obvious. He apparently roomed alone and there is no record of having either male chums from outside the regimented Kummersdorf-Peenemünde coterie to socialize with or female companions other than secretaries and occasional girlfriends of no consequence. He went deer and bird hunting with Dornberger. While waiting for Peenemünde's labs and test stands to be ready, he turned part of his attention to the joint Luftwaffe–army project to develop rocket-powered fighter aircraft that had been instigated by von Richthofen.[28] Despite his personal involvement, which included climbing into a Heinkel 112's cockpit to switch on its experimental Kummersdorf engine to impress a Luftwaffe test pilot, the configuration resulted in setting the plane's tail on fire during an abortive flight in June 1937. A different engine was ordered from the Heinkel company, rather than Kummersdorf, and the project continued on another track. By then, Göring had moved von Richthofen aside for a more famous Great War pilot, Ernst Udet, whereupon von Richthofen found his own brand of glory over Guernica.

THE OVERALL PICTURE for von Braun at this interlude, between the first four years of his work for the army in Berlin and Kummersdorf and the eight he would spend at Peenemünde, was a privileged position within a military establishment devoted to Hitler and driven to prepare for war in the near term. Money was flowing into rocket work as never before, but skepticism about what it could contribute on the battlefield was high right up to the führer himself, and rightfully so. If von Braun wanted to maintain his status, he needed to subscribe loyally to official policies that, for example, made hiring anyone who could not present an *Ariernachweis* (the Certificate of Descent, or "Aryan certificate," based on baptism records for parents and grandparents) illegal. A February 1934 decree had already made so-called Aryan descent necessary for service in the Wehrmacht. After April 1937, Jews were no longer allowed to earn degrees in any subject from German universities.[29] All personnel brought into the Peenemünde fold therefore passed through a racist filter, which was wielded by von Braun whether he contemplated it or not. There is no evidence that he ever questioned these conditions, let alone resisted them.

"I confess to no deep psychological thinking on this matter during these times," he wrote decades later in a rare rumination, adding that "I did not have any more scruples in accepting [army] support than, say, the Wright Brothers may have had when they signed their first contract with the U.S. War Department."[30] That he found the two circumstances comparable lends considerable credence to his claim of never having thought about it.

The technical order of business before him was to help fire the A-3 rockets successfully and proceed apace to real weaponry. Because the test stands at Peenemünde were still under construction, Dornberger ordered the launches to take place from the Greifswalder Oie, a tiny island some 5 miles off the northern tip of Usedom with a lighthouse and rudimentary fishing harbor. It had been leased by the city of Greifswald to farmers over the years, one of whom ran a humble Inselhof (inn) that attended to "the warmth of the outer and inner man," as Dornberger remembered. There was money now for a much more ambitious operation than the launching of little "Max" and "Moritz" from Borkum three years earlier, and the army's obsession with secrecy evidently outweighed the logistical drudgery of transporting men and gear across open water to a primitive site as winter approached. The rockets would be set directly upon their tail fins, fired straight up, and returned by parachute—a sequence that would someday become familiar to almost every man, woman, and child on the planet. During these development stages, at least, having them come back down anywhere but into the sea, with their precious and secret contents, was unthinkable.

8

GRAND AND HORRIBLY WRONG

L OOKING BACK ON the early days at Peenemünde, on the expedition to the Greifswalder Oie and the challenge of finally building big rockets that were not toys, Wernher von Braun might have wondered in 1943 whether it had all been a dream. In many ways, it was. But no record of his thoughts is known to exist, no diary, journal, or personal letters that could shed light on how he felt about the whole momentous scene. For a man from the highest cultured stratum of German society, this is a most peculiar void. After 1945, he would quickly become a loquacious author of popular books and articles, but before that terrible line in history it would seem that he wrote nothing except official desk traffic. Perhaps he had no private thoughts worth putting down or time to write them. Perhaps military censorship trampled the urge. Perhaps he lost his personal missives in the chaos of war. Perhaps they disappeared the same way his swastika lapel pin and SS uniform did. In any case, historians were left with a preternaturally one-dimensional record, reasonably construed as a sanitized record, which when read in the academic historiographic tradition yields a nearly one-dimensional man. This figure, the "dreamer of space," still strides across the landscape three decades after his death, a testament to how closely it matched a technological society's desire for such a spirit.[1]

By 1943, German rocketry was in the clutches of something horribly

96

wrong, and any connection to the whimsical days of the Raketenflugplatz was an exercise in nostalgia. The hale comradery of Borkum and the Greifswalder Oie had given way to a grisly fight for national survival, shadowed by genocide. Similar violence could have been experienced by von Braun's Prussian ancestors during the Napoleonic wars or by his parents' generation in the Great War, but his attempts later in life to universalize his role as an armorer, to compare himself to the Wright brothers or the Manhattan Project scientists, so long after the singular crimes of the Third Reich had become widely known, stand as the starkest measure of the quality of his moral thinking.

On May 1, 1937, von Braun became a member of the Nazi Party, number 5,738,692.[2] While the Wehrmacht had maintained a Reichswehr regulation against political affiliations to control the influence within its ranks of paramilitary groups, von Braun as a civilian employee was not subject to it. Ten years later he would testify that he was "officially demanded to join," but offered no explanation of who applied such pressure. As in his post hoc account of joining the SS, he claimed that "my refusal to join the party would have meant that I would have had to abandon the work of my life," but he did not name Dornberger or anyone else as the source of this career guidance.[3] He also stated that he joined in 1939, a two-year error that went unclarified. At the very least, according to his own words, it was another example of placing "work" above all other considerations. His work, of course, was developing weapons for the Hitler war machine.

At the end of November 1937, when four A-3 rockets packed in dark gray boxes arrived on the Greifswalder Oie aboard a decrepit nineteenth-century ferryboat (no doubt for the sake of secrecy rather than a whimsical evocation of when steamers once brought day-trippers from Wolgast), von Braun's work was about to meet its stiffest test yet. Since the previous spring, laborers and engineers had done everything from dredging the harbor for cargo vessels to installing sophisticated instrumentation on the 54-acre undeveloped island. They got bogged down in the roadless Oie's wet clay and felt lucky to make use of narrow-gauge railroad tracks left after the lighthouse was completed in 1855. Baltic storms played repeated havoc with the preparations, especially as winter approached. They might easily have felt like cast members in the UFA science-fictional movie *FP1 antwortet nicht* ("FP1 Is Not Responding"), filmed on the island in 1932, about a floating platform in mid-ocean for servicing transatlantic flights that is wracked by storms. During the summer, von Braun had gained use of an army biplane to fly between Peenemünde and Berlin, though this would not have saved much time in an era when facilities for such travel were still in their infancy. Rather like Hitler's 1932 election campaign flights, von Braun's barnstorming lent a bit of dash to the army's staid reputation.

As the launch date for the first A-3 approached, about 120 men gathered to assist or observe. It was named *Deutschland*—certainly with the bombing of the pocket battleship *Deutschland* by Spanish Republicans at the end of May in mind, which had killed more than a score of sailors— instead of something playful like "Max" and "Moritz" or science-fictional like "Repulsor."[4] While the technicians fretted over last-minute glitches that would come to typify the technology, von Braun and Dornberger whiled away their time shooting the Oie's abundance of pheasants and rabbits. During the region's centuries of feudalism, such recreation would have been the prerogative of the Duke of Wolgast and forbidden to commoners. The fact that von Braun's grandfather was said to have hunted ducks around Peenemünde pointed to the family's aristocratic heritage.[5] Soon a jaunty pheasant plume became the expedition's ubiquitous hat decoration. As a symbol of von Braun's status as young Freiherr and management executive, the Greifswalder feather was right on target.[6]

After a year of concerted effort and a leap of magnitude in personnel to about 400, the A-3 launches represented the first fruit of heavy investment by the Third Reich in military rocketry, as well as the debut of operations in the Peenemünde zone. All eyes must have been focused on the occasion, at least those permitted to know about it. On December 4, a boatload of "dignitaries" bobbing in a cold swell waited impatiently offshore while the first shot was delayed for hours by electrical problems. The rocket had been painted with water-soluble green dye to aid recovery from the steely Baltic, but it was dissolved by condensation from the −297°F liquid oxygen inside and dripped down the fins, shorting various cables at the base.[7]

Finally, after three years of labor to evolve the A-2 and many blue-sky promises made to purse-string holders, "the ignition functioned to perfection and the first A-3 began to rise majestically towards the zenith," von Braun remembered. But after just a few seconds, its recovery parachute sprang out by surprise, "streaming into the fiery jet which consumed it in an instant." *Deutschland* spun, tumbled, and promptly crashed in a fireball of unspent propellants.

Whether the dignitaries returned for the second attempt is unknown, but they had already seen everything the A-3 could do. One after another, in the span of a week, the rockets rose from their four-legged metal firing table for a few seconds and then fell—three out of four into the sea. Retrieved fragments told nothing about what went wrong, but observers —including Dornberger up in the 160-foot-high lighthouse—noticed that with or without the parachute (it was removed after the first two failures), the rockets spun lengthwise like a drill and turned into the wind, instead of shooting straight up with no roll as the Kreiselgeräte control system was supposed to ensure. At an altitude of only several hundred meters, they all tumbled over and dropped from the sky.[8]

The despondent engineering staff realized that Kreiselgeräte's mechanism was the central problem, exacerbated by immature understanding of the A-3's aerodynamics. Von Braun, who had made the initial contact with Johannes Boykow, demonstrated for himself on a wooden box how the spinning gyroscopes could cause the horizontal plate they were mounted on to flip if it rotated around its vertical axis too fast, a completely unforeseen possibility.[9] The Kreiselgeräte device could control only the rocket's pitch and yaw, not its roll. And fresh calculations showed that the steering vanes in the exhaust jet could not counterbalance the A-3's tendency to roll in the first place. The pesky parachutes had actually done what they were supposed to do—deploy when the rockets tipped over, which would normally have occurred at the peak of their trajectory after engine cutoff. All the work had thus resulted in a fundamentally flawed rocket that had looked good on paper and during partial tests, but was doomed under real conditions. If von Braun was the gifted systems engineer whom everyone later praised, it would have been up to him to understand before committing to full-scale launches that the A-3 control system was no match for A-3 aerodynamic forces. Boykow and Hermann had done what they were paid to do in their respective specialties, but they were not responsible for integrating the two.

Many years later, von Braun tried to shed blame on the long-dead Boykow, claiming that many of the gyroscope expert's assistants had disagreed with their boss.[10] But this suggested that as a manager he had either trusted Boykow's expertise too much for such novel technology or allowed himself to be poorly informed by a crucial contractor. His youth was sorely exposed.

The 1937 holiday season could not have been as joyful as von Braun and Dornberger might have hoped. After weeks of meetings in the new Designing House at Peenemünde-East, knowing that it would take at least a year and a half to prepare a better control system for another launch, Dornberger decided to abandon the A-3 instead of tinkering with it. Because the next sequential Aggregate number, A-4, was already in use for the big long-range missile that was still on the drawing boards, the improved device would be designated A-5, to avoid the "stigma" of the A-3, as von Braun recalled.[11] Four spectacular crashes had obviously poisoned the option of calling it anything like A-3b. In a military culture universally obsessed with such numberings, "5" would also sound like a major advance from "3," despite what had clearly been a sickening embarrassment on the Greifswalder Oie.

The experience completely changed Dornberger's technical modus operandi, as well. To avoid long periods of time between launching one Aggregate series and the next, which naturally created huge expectations, and to acquire knowledge in smaller experimental steps so that so much work would not have to be scrapped again, he ordered the production of ten A-5's per month. The army could afford it now, and he really had no choice but to run many more test flights, given the growing complexity of the technology.

WHAT NEITHER DORNBERGER nor anyone else could know was that the clock had all but run out on German rocketry's halcyon era, if any time after 1933 can be called that. The year 1938 marked the Third Reich's direct and reckless turn into the teeth of war. In early November, while Peenemünde technicians were still preparing the Oie for A-3 launches, Hitler had used the occasion of a meeting of military chiefs of staff ostensibly called together to discuss allocations of raw materials to reveal his bellicose thinking about foreign policy, in particular its pell-mell timetable. *Lebensraum*, his racist vision of "living space" for a German empire through territorial expansion in Europe, could only be obtained by force—and soon, meaning 1943–45 at the latest. The superiority of German armaments would last only that long, he believed. Attacking Czechoslovakia and Austria might be necessary as early as 1938. While the military leaders did not disagree with the broad aims of Lebensraum, they were stunned by the direct implication that Germany might find itself at war with France and Great Britain in the very near future. For most of them, after all, the Great War would have seemed like yesterday. General von Fritsch, who had opened the flood gates of funding for Dornberger, was so disturbed that he needed to be assured by Hitler that war was not so close as to require him to cancel a planned leave.[12]

In January, while the Peenemünde engineers were puzzling out how to proceed after the A-3 debacle, the top military leadership was rocked by two scandals that resulted in Hitler's seizing complete control of a Wehrmacht that seemed to him reluctant to back his *Auftrag*, or historic mission. Werner von Blomberg—a führer-infatuated general who took over from Kurt von Schleicher as minister of defense when Hitler became chancellor—had recently married a woman thirty-five years his junior, a stenotypist whom he had met during a stroll in the Tiergarten. Hitler responded personally to the field marshall's quaint concern about taking a socially inferior wife by serving as witness at the wedding, along with Göring. When it was discovered within days that the bride was in fact a prostitute who had also once posed for pornographic photos taken by her Jewish boyfriend, the regime faced a crisis exquisitely sharpened by its own ideological obsessions. No sooner was Blomberg dispatched to comfortable exile in Italy, than General von Fritsch, whom Hitler was considering as Blomberg's replacement at the War Ministry, became embroiled in an old allegation of homosexuality.[13]

While fretting over the Blomberg scandal, Hitler had recalled that in 1936 Himmler had passed to him a confidential file about suspicions that von Fritsch was being blackmailed by a male prostitute. Himmler promptly relocated the file, though Hitler had ordered him to destroy it. Confronted with the charge, von Fritsch angrily denied it, which prompted a bizarre meeting at the Reich Chancellery with Hitler, Göring (who wanted for himself the vacant Defense Ministry position), and the supposed blackmailer who was fetched from a concentration camp in northwest Germany for criminals (which included homosexuals). Von Fritsch gave his word of honor that he had never seen the man before, but Hitler was not convinced. Prodded by an investigative report on the matter written by Justice Minister Franz Gürtner—the von Braun family friend—who stood legal tradition on its head by opining that von Fritsch had not proven his innocence, Hitler demanded the general's resignation.[14]

To mask the sudden hatcheting of two of the nation's highest military leaders, Hitler quickly set in motion a sea-change of personalities at the top of the Wehrmacht, Luftwaffe, and diplomatic corps that finalized his absolute control over the instruments of strategic power. Newspapers whipped up a torrent of public speculation about imminent war. The army was most affected, losing the last vestiges of traditional independence from the Nazi Party and SS.[15] "In this Reich, everyone in any responsible position is a National Socialist," Hitler declared in one of his harangues before the Reichstag in February.[16] The Blomberg-Fritsch imbroglio may or may not have been shaped for the purpose of consolidating Hitler's control over military and state, but it delivered total dominance into the hands of the supremely opportunistic führer. At just the moment when the radical nature of his ambition was coming to the fore, the old conservative establishment that might have tempered it was emasculated and pulled under the führer principle.

IT WOULD TAKE until October 1939 for the Peenemünde engineers to achieve with the A-5 what they had failed to do with the A-3 in December 1937. During those twenty-two months, their universe changed forever, and not for the better. Whatever admiration Hitler had garnered from a world willing to look past his domestic infamies to the material rebirth of Germany was about to be dashed along with the brief golden age of Aufschwung. As an R&D program at the cutting edge of a new technology, the rocket work was moving forward as fast as anyone could reasonably expect. But the supply of reason rapidly dwindled until it was gone, leaving von Braun, Peenemünde, and all of Germany stranded in a nightmare from which no one could awaken.

Von Braun's primary professional concern during this time was the control system that had caused such a deep setback on the Greifswalder Oie. Until the problem of how to keep rockets on course was solved, they were useless, especially as long-range weapons. While Walter Thiel's group at Kummersdorf devised engine innovations that would feed directly into the A-4, Dornberger reached out from under his preferred army *Dach* to contract with commercial companies for new gyroscopic gear. Thiel had essentially displaced von Braun from engine design, the subject of his doctoral dissertation and first practical experience. Aerodynamics was under the firm direction of Rudolph Hermann. This left guidance and control, the third major segment of rocket development, a field in which von Braun possessed no theoretical or design expertise. His role as a supervisor was thus cemented as he coordinated efforts by Kreiselgeräte, Siemens, and other companies. Eventually, a fully staffed laboratory arose at Peenemünde for what engineers call "integration and testing" of components supplied by industry, a process notorious for taking more time than estimated.

One expert whom von Braun hired for in-house guidance and control work would resurface many years later as the only Peenemünder to criticize him harshly and openly. Dr. Paul Schröder was a mathematician with whom von Braun had "continuous collisions" after he was brought in during 1937 to head the bureau developing equations to describe the stabilization of rockets in flight. According to Arthur Rudolph, von Braun found Schröder to be too cautious or pessimistic about the technical challenges ahead.[17] Rudolph also recalled that Schröder had predicted the A-3 failures with his theoretical calculations, which would have been a good reason for him to be cautious. The relationship reached the breaking point and von

Braun told Dornberger that Schröder was no longer needed. Among Schröder's accusations was that von Braun "ran roughshod over scientists at Peenemünde."[18] Dornberger did not fire the talented theoretician, but moved him first away from von Braun's sphere, then into redundancy, and finally to nonrocket work for Army Ordnance. Schröder was later described elsewhere as accusing von Braun of being "an opportunist who picked the brains of better men and grabbed credit others should have had."[19]

Arthur Rudolph also later alluded to von Braun's penchant for barging onto the turf of other Peenemünde engineers. During 1937 and 1938, when organizational charts were being established, Rudolph experienced "frictions" with a von Braun "bubbling over with ideas" who would enter Rudolph's workshops without consultation and try to make willy-nilly changes. Von Braun was so manic in this regard that "Even Dornberger had to say [to von Braun], 'No, Godammit, stick to this, this, this!' " The military leader of the rocket program "always saw to it that the ever bubbling ideas from von Braun did not mess things up."[20] Dornberger generally concurred in his memoirs, recalling that von Braun could be erratic, growing stubborn and intolerant when he decided what he wanted to do.[21] Yet Dornberger always seemed willing to clear obstacles from von Braun's path in a fatherly way.

There is thus an alternative picture to von Braun as the genius maestro of Peenemünde who knew more than everyone else: the young turk, coddled by a paternalistic Dornberger, who was given little actual hands-on work to do and kept busy by putting his nose into everyone's business. Dornberger insisted that they never quarreled or had differences of opinion that could not be resolved, that he would do anything to smooth von Braun's way. These were extraordinary terms for such a setting. Though Dornberger praised von Braun as an "indisputable genius," as did so many others who lived to talk about Peenemünde, the image left for historians can undergo gestalt shifts like the drawing of ugly witch or beautiful lady.

WHATEVER PERSONALITY PROBLEMS arose (and they would of course be expected in a program that brought together in close quarters so many bright young minds), the prime lubricant in the rocket work was money, which Dornberger had remarkable success in obtaining. Hitler's replacement for General von Fritsch as head of the army was Walther von Brauchitsch, a Junker artillery veteran from the old aristocratic military caste, whose professional association with Dornberger traced back through the Weimar years. Moreover, he was an alumnus of the French Gymnasium in Berlin, von Braun's alma mater, and had married an heiress to a 300,000-acre estate in Pomerania. In addition, as part of the turnover in numerous commands during von Fritsch's departure and von Brauchitsch's arrival, Karl Becker became chief of Army Ordnance. Dornberger could hardly have enjoyed a more favorable alignment of funding stars for Peenemünde, though delivery of an actual weapon from its gates still waited somewhere in the nebulous future.[22]

There was only one star missing, but it happened to be the one that mattered most. Without the führer's personal imprimatur, the rocket program would find itself jostled among myriad armament initiatives, all with high priority within their respective services, competing for resources that were stretched to the limit and beyond. Dornberger, Becker, and von Brauchitsch could not yet know how crippling this byzantine procurement contest—which placed Hitler at the fulcrum of all decisions—would become when war finally broke out. For now, they understood that winning his keen interest was crucial, and that there was no better way to accomplish this than by firing for his personal edification one of their thunderous rocket engines, which had rendered awestruck so many important men before.

They were compelled to take such action not just by the extraordinary level of investment in men and material they knew would be necessary at Peenemünde, but by Hitler's turn toward aggressive military force in the spring of 1938. In March, German troops entered Austria and the country was promptly annexed as a province of the Reich. Though supported by jubilant Austrians, whose small German-speaking nation had been fabricated as a remnant of empire in 1919 by the Great War's victors, *Anschluss* (union) was not simply a fulfillment of pan-Germanic self-determination but a bullying chess move by Hitler to improve his strategic position in Europe.[23] That the Wehrmacht was completely unprepared for the invasion, which fortunately for the generals met with no opposition either on the ground in Austria or from other European governments, would underline for von Brauchitsch that Hitler was the only man in charge now.[24] Most ominously, the successful maneuver unleashed a fury of terror against Jews and leftists in Austria, organized by Himmler, that spilled back into Germany during the summer of 1938. From here on there could be no doubt whatsoever about the Third Reich's inherent malevolence.

The next move followed quickly. In April, Hitler ordered the army to plan for an invasion of Czechoslovakia, which Germany now encompassed since the addition of Austrian territory. This offensive clearly went beyond the nationalist program that had been generally supported since 1933 by political, economic, and military interests in Germany. Once again, the army was shaken by the prospect of a war that might draw in Czechoslovakia's protectors among the European powers, especially France and the Soviet Union. But British and French officials made clear that they would not intervene if Germany attacked. Somewhat ambiguously, British Prime Minister Neville Chamberlain warned Hitler against the use of armed force, but left open the possibility of negotiated revisions of the Versailles settlement that had created polyglot Czechoslovakia in 1918. This, of course, would have been anathema to Hitler, who beheld the Czechs as an inferior race and had sworn to smash the diktat from the beginning of his political career.

At a mid-September meeting between Chamberlain and Hitler at Berchtesgaden, the prime minister offered to cede Sudetenland, the ethnic German region in western Czechoslovakia. Hitler called his hand, however, by demanding an immediate German occupation of that territory, while lying that this would be his final expansion in Europe. The demand caused a war panic, with Londoners waiting in line for hours to be fitted for gas masks. At the end of the month, at a conference in Munich attended by Hitler, Mussolini, Chamberlain, and French Prime Minister Édouard Daladier—but no Czech (or Soviet) leader—Hitler was granted the Sudetenland and de facto control over the rest of the country, which remained nominally independent. Chamberlain returned home from the carving up of Czechoslovakia to announce that he was "bringing peace with honor—I believe it is peace for our time." Some prominent European diplomats, on the other hand, thought that Hitler was going insane, and a coterie of high-ranking Wehrmacht officers, possibly including von Brauchitsch, were planning a coup d'état if Hitler ordered an invasion.[25]

AGAINST THIS TENSE backdrop, when the threat of war was palpable all across Europe, A-5 rockets began to be launched from the Greifswalder Oie in October 1938. A new guidance and control system was still unavailable, so the trials occurred with as little crosswind as possible and merely tested the inherent stability of the rocket body itself. Only two were fired off, several days after the Munich agreement. "The thunderous roar of its jet smote the ears of the listeners for upwards of a minute," von Braun grandly recalled, followed by "cries of joy." It was something to cheer about at a moment when continued silence from Peenemünde was clearly untenable as Hitler rattled his sabers. On the night of November 9, the pogrom that would come to be called *Kristallnacht*—during which scores of Jews were murdered and tens of thousands sent to concentration camps, causing international outrage (the United States recalled its ambassador from Berlin in protest)—added a scream of its own to the atmosphere.

Events in the world were moving fast even if progress at Peenemünde was not. On March 16, 1939, after the German army marched unmolested across the Czech border, Hitler appeared in Prague to announce the formation of a German protectorate. He returned triumphant to Berlin on March 19 to enjoy one of Goebbels's trademark searchlight-and-fireworks mob spectacles. So-called appeasement policy lay exposed as a pipe dream of governments still weary from the Great War. Britain and France now realized they had no choice but to speed up their own rearmament.

On or about Wernher von Braun's twenty-seventh birthday, March 23, von Brauchitsch and Becker brought the führer on a raw rainy day to the old Kummersdorf proving grounds, where he witnessed two static firings of rocket motors. They were no doubt confident that the thunder would work its usual magic. Getting him all the way up to the Baltic coast was evidently too much to arrange—indeed, the führer would never visit the utopia of German R&D. Dornberger remembered long afterward that

Hitler's mind seemed to be somewhere else—as well it might have been, given the fact that he had just finished rearranging the map of Europe. Von Braun, meeting Hitler personally for the first time, gave his well-practiced spiel about the new technology (minus any blather about space travel, which Dornberger specifically prohibited him from mentioning). Hitler left the demonstration shaking his head. He consumed his usual vegetarian lunch and glass of Fachingen mineral water, asking Dornberger how long it would take to develop the long-range A-4, which might actually have some military value. The führer then "looked past me with an absent smile," as Dornberger recalled, and said, "Es war doch gewaltig!" Well, it was grand.[26]

And there it remained, very grand, but useless, five months before the start of World War II.

9

DEPRAVITY

Wernher von Braun's encounter with a deeply distracted Hitler five months before the invasion of Poland contrasted with the focused attention of Himmler seven months later, when the Reichsführer-SS dispatched a minion to Peenemünde to secure his membership. As a reflection of the Third Reich's polycentric leadership, this experience should have crystallized for the young man the multiple poles of the totalitarian dyad now steering his life. Hitler was a mercurial demigod, convinced of infallibility. Supremely successful conquests of Poland and France pushed him deeper into megalomania. Himmler was singularly criminal, but lacked the charisma to connect himself to anything much wider than his own sphere of power.[1] The murder and enslavement he directed in the wake of Hitler's invasions would only make him more desperate to save his own neck after the disastrous attack on Russia, Operation Barbarossa. In a closed universe where any notable individual had to make an in-or-out deal with Hitler and Himmler, von Braun always chose *in,* claiming later that it was for the sake of his "work."

With war came mobilization, which threatened to drain away the talent that Dornberger and von Braun had so meticulously collected at Peenemünde. If there had been no war, no fanatical Lebensraum, no pathological wish to exterminate the Jews, no paranoid fear of Bolshevism—no

Third Reich, in other words—then perhaps the grand experimental station on the Baltic coast would have continued along its methodical path toward the rockets von Braun had dreamed of as a child. It could have done this, given the truly undisputed genius of pre-1933 German science and technology. But the Peenemünde utopia and the Nazi dystopia comprised a single, split personality. Peenemünde was the Third Reich. They formed a compound of science and society that could not be separated, except perhaps in the minds of men who believed that science occurs in a politically antiseptic realm, who later wished to erase the past after the devil had taken his share.

For the Peenemünders, the war divided into two phases. From September 1939 to August 1943, they faced the stark reality under wartime exigencies of weaponizing experimental rockets that were still essentially flying laboratories. Although war is a well-known accelerator of technological progress, it is a single-minded and rapacious master that can quickly burn through enormous resources. The technical problem of building serviceable long-range guided missiles was thus complicated by cutthroat infighting for money, manpower, and materials in a byzantine procurement system that pivoted on a single point: a führer increasingly detached from reality. Blasé about rockets at the war's start, though appreciative of their revolutionary potential, Hitler badly needed "wonder weapons" three years later. Production facilities at Peenemünde had been conceived for turning out modest numbers of test missiles, not for the thousands demanded by tactical deployment. The test missiles themselves underwent constant modification, sometimes ad hoc and not always closely codified, which was anathema to mass production. Missing from the engineering staff were experts in converting from development to manufacturing. Formal drawings of components, which were necessary to build anything in great repetition, were often scarce or nonexistent.[2] In this regard, Dornberger's "all under one roof" organizational scheme, which he later claimed was born from the need for secrecy but also created a compact empire, became a serious handicap. Little was in place for transferring even a technically mature device to industrial-style serial production—and the rockets were never mature.[3]

Then, from shortly after the Royal Air Force bombing of Usedom in August 1943 to the end of the war in April 1945, Peenemünde supported the subterranean assembly of missiles by slave laborers under SS rule at Mittelbau-Dora. It was during this period that von Braun visited the Harz Mountain complex at least fifteen times to coordinate the effort.[4] Nazi Germany was now fighting for survival, not for a thousand years of glory, and the most fanatical figures—particularly Himmler—took charge. In case of defeat and unconditional surrender—which the Allies called for at the Casablanca Conference in January 1943—courts would seize upon the highest-ranking individuals in the regular army, which generally adhered to international law. But SS atrocities in occupied territories were widely known and many of its members throughout the ranks would be considered criminal by definition.[5] This is why von Braun's SS rank, albeit "honorary," and his wearing of the black uniform would be suppressed after the war, when the world was in no mood to weigh fine gradations in SS affiliations.[6] Albert Speer recalled many years later that "during the Nuremberg Trial, I mused that such an honorary title would have destroyed any chance of my survival."[7]

Putting aside the clear lens of hindsight, it is still reasonable to posit that the second phase of the wartime rocket program grew inevitably from the first. That is, the technical problems of making a weapon out of an experimental device led the program into desperate measures when production became an absolute necessity. The obvious vulnerability of Peenemünde to air attack from the spring of 1942 (when the RAF began its offensive against Germany) onward, the infighting for resources, and the hell-bent measures that were necessary for mass production all combined to turn the V-2 into a monster child. After the coup d'état assassination attempt by army officers on Hitler of July 20, 1944, Himmler ruled at will and Dornberger, a practical man, fell in line. "Had the SS not taken over the development of the rocket, it never would have happened," a key Peenemünder admitted years later.[8]

The "paternalistic" Dornberger—this was Albert Speer's adjective for him—did everything he could to keep the money flowing, but the ground kept shifting under him, as it always does during war.[9] His workaday concerns turned more and more to matters of using the missiles rather than perfecting them, while his struggles with the SS over control of the program grew acute. Yet it was Dornberger himself who finally opened Peenemünde's door to the Reichsführer-SS, when in December 1942 he sought Himmler's help—via Gottlob Berger, a Himmler confidant and SS headquarters chief—in arranging an audience with Hitler for himself and von Braun to pitch the rocket program.[10] Dornberger's success in protecting Peenemünde bureaucratically for the first six years of its existence, enhanced by the center's geographic remoteness from Berlin, was a prime factor behind the rocket program's technical accomplishments. After that, the dogs of war took hold. It was, after all, a weapons program—even Peenemünde's architectural splendor began to grate against wartime sensibilities.[11]

What never changed was the artillerist concept of a missile that could carry a ton of explosives (the actual payload turned out to be only 1650 pounds) to targets over the horizon, which Dornberger and von Braun had sketched out in the salad days of 1936. But by autumn 1944, when the missiles were finally ready, fleets of British and American bombers were pummeling German cities and military targets with thousands of tons of bombs per day. It was a situation that could not have been foreseen in the mid-thirties—long-range heavy bombers were still in their infancy then, no one would have imagined the loss of air supremacy over the German homeland, and even the 1-ton rocket payload entailed a great technological stretch. Speer was especially caustic after the war about this imbalance. "The whole notion was absurd," he wrote of the missiles. "Our most expensive project was also our most foolish one."[12] On a cost-benefit basis, the investment of billions of marks in missiles might have been justified militarily if they were either available in vast numbers (as Hitler, out of touch with technical realities, wanted) or could carry explosive payloads comparable to heavy bombers. Neither was ever the case. Indeed, the latter capability would not come of age for another generation, until the advent of lightweight atomic warheads.

Peenemünde was thus hoist by its own petard. Dornberger sustained the budget of his pet program from the beginning by conjuring a Fata

Morgana on optimistic deadlines for doubtful superiors. Before the realities of the Russian front and American entry into the war became impossible to ignore, he and von Braun even floated the Nebelesque possibility of building multistage intercontinental ballistic missiles, which were wildly beyond state of the art.[13] Hitler was the most skeptical of all, having downgraded the rocket program's priority for labor and materials right after the war began, in November 1939, to prepare for the invasion of France in the midst of a munitions shortage following the Polish campaign. His refusal to approve mass production until a long-range missile had successfully flown was clearheaded wisdom compared to the constant bugling of Peenemünde's leaders. Both Dornberger and Speer faulted him many years later for lack of imagination regarding the rockets, for being stuck in the technical experience of the Great War's trenches, but the führer obviously understood that invasions still required massive quantities of dependable conventional weapons. Even the program's godfather, Karl Becker, seemed to be losing faith when, two days before he shot himself in April 1940, he said to Dornberger, "I only hope that I have not been mistaken in my estimate of you and your work."[14] Though neither Dornberger nor other army leaders foresaw that total war would come as soon as it did, or that they would lose so much autonomy in making procurement decisions, he and von Braun always knew that the rocket's development timeline stretched off into the hazy future. And they were perfectly aware from the start that the long-range missile would pack a relatively modest punch, because they had set the 1-ton payload parameter themselves.

It is safe to assume that army officials tolerated the rocket program because it became big and rich, a time-honored rationale in military-industrial establishments that has far more to do with careers than with battlefields. It is therefore not to Dornberger, von Braun, or any other Peenemünder that historians must turn for some deeper understanding of why the missile project rolled forward year after year, but to the hauntingly eloquent Speer. The rocket work "exerted a strange fascination upon me," he remembered. "It was like the planning of a miracle," which is exactly what the Reich eventually needed. This almost spiritual impact of

the powerful new technology, analogous to the emotional effect of wit-
nessing heavier-than-air flying machines for the first time a generation
earlier, is difficult to conjure up seven decades later. From the winter of
1939–40, when he began to oversee Peenemünde's construction needs and
skirted Hitler's cutbacks, Speer was "impressed anew by these technicians
with their fantastic visions, these mathematical romantics." Whenever he
visited the seaside center, he felt "somehow akin to them."[15] Here was a
succinct expression of one of the conundrums of Germany's Nazi era, that
in the midst of sociopathy, educated and cultured individuals could find
inspiration, or at least wish that they could find it. Speer was a schemer of
great category, of course, and his remarks must be parsed with care for
deviousness.[16] But he was unique among the Third Reich's major figures,
and certainly among the men who later wrote about the rocket program,
in confronting large issues of conscience through his postwar autobio-
graphical writing, whether speciously or not. By comparison, von Braun's
recorded self-reflections after the war were shallow or obfuscatory.

Like many of his countrymen, Speer seemed to yearn for something
positive to come out of the time, after Germany's years of humiliation and
misery following the Great War, which meant being able to turn a blind
eye to the irreparably negative. He wrote about how he "liked mingling
with this circle of non-political young scientists and inventors," as though
working at a German weapons laboratory after 1933 were a neutral act.
After the war, he struggled to understand how his moral calculus could
have been so wrong, whereas von Braun never did to any significant
known degree, relying instead on the juvenile excuse—accepted by many
Americans to this day—that space travel was what he was really working
on all along. Speer was far more sophisticated than this, yet what they
appear to have shared was a belief, faith, or stupefying desire that German
genius could somehow shine through the ferocious modern age. Here was
Thomas Mann's "highly technological romanticism" in action, the Zeitgeist of Jünger and Spengler come alive.

None of the Technik und Kultur intellectuals divined what was to happen at Mittelbau-Dora, of course. If anyone had, it is easy to imagine that
they would have been considered insane. Himmler's longstanding ambi-

tion to create autonomous armament plants under SS management took hold of the rocket program within days after the RAF ruined Peenemünde and just a month after the failure of Hitler's last attempt to stage a breakthrough on the Russian front at Kursk. Himmler's plan to force concentration camp prisoners to build weapons had been approved in March 1942 by Speer's office, with a carbine factory at Buchenwald serving as a prototype.[17] It was a stark technocratic decision unmarked by humanitarian considerations, one that underscores the incredible depravity to which the Third Reich was sinking. Speer later maintained that both the army and commercial industry opposed establishing concentration camp factories because they did not want to go up against the SS, but the wartime shortage of conventional labor made the choice look coldly logical. By his own account, the subsequent manufacturing centers were grossly inefficient, unprofitable, and plagued by murderous living conditions even after Himmler himself ordered in December 1942 that mortality rates had to be reduced for the sake of output.[18] Mittelbau-Dora exemplified all of these problems to the extreme, particularly the human toll.

Von Braun was already well acquainted with slave labor when construction began at the underground site in the late summer of 1943. He had been living and working beside it for a year with no objection. Germany had neither the population nor the domestic resources to win a war against the United States and the Soviet Union, hence Speer's approval of *Zwangsarbeit*—forced labor of prisoners of war, civilians from occupied territories, and concentration camp inmates that included German criminals and political detainees. A non-SS contingent of thousands of Polish and Soviet workers arrived at Peenemünde in 1942 to build infrastructure and huge facilities such as the electrical generating station, liquid-oxygen plant, and missile production hall. They were impounded at Trassenheide, a village on the road between Karlshagen and Zinnowitz. The Buchenwald concentration camp near Weimar supplied laborers, which also included French and German *Häftlinge* (detainees). On July 7, 1943, for example, a contingent of several hundred Frenchmen was loaded into trucks bound for Weimar and then by train to Peenemünde, where it joined about 400 French, Russian, Belgian, Dutch, and some 50 German prisoners in a single large barrack surrounded by electrified barbed wire and patrolled by about fifty SS guards.[19] They rose at 5 A.M., worked from 7 A.M. to noon and 1 P.M. to 7 P.M., had Sundays off, and were fed coffee for breakfast, soup with bread and margarine for lunch, and soup again for dinner. Discipline was "rigoureuse, sans plus," with the only real danger being two violent SS guards nicknamed "Moustache" and "le Hibou" (the Owl). This relatively benign existence changed drastically after the RAF raid in mid-August, which killed some 600 prisoners. Living conditions deteriorated and those who survived bore the vengeful brutality of the German-born "kapo" inmates, often hardened criminals whom the SS used as police.

Von Braun was not responsible in his official capacity as technical director for initiating forced labor, which required Dornberger's nod, but he was certainly more than just aware of it. One of his closest associates, Arthur Rudolph, advocated the use of SS prisoners for missile production and then directly managed it.[20] The workers were not forced on Peenemünde by the SS, as Dornberger and von Braun would portray years later. In the summer of 1943, slaves began to install missile production equipment at Peenemünde-East.

Concentration camps for Hitler's political opponents had been an open fact of life in Germany for ten years, so there was little or no social opprobrium left against the existence of throngs of prisoners who were by now from myriad walks of life, military and civilian, all ensnared in the SS's sadistic, racist system. If von Braun had any qualms about slave labor, they were technocratic. No pre-1945 record exists of any other kind of consideration among the Peenemünde leadership.[21] After the war, when self-justification was rife, claims would be made by various principals that they had tried to aid the prisoners, but under the circumstances there was only one impetus to do this: to improve production output. As long as the workers were slaves, any improvement in their personal lot carried the potential for increasing the availability of weapons for German forces trying to subjugate all of Europe. They were quite aware of this Kafkaesque twist in their hellish existence. In one case that has been much discussed over the years, a French professor, Charles Sadron, who was enslaved at Mittelbau-Dora after being arrested for *Résistance* activities, bravely refused von Braun's personal invitation to work in his own comfortable laboratory, rather than be drawn into collaboration with the Nazi war machine.[22]

THE IMMEDIATE PUSH for moving from the Baltic coast to the Harz Mountains was the devastating RAF raid, which did not destroy Peenemünde's production facilities, but made it clear that the Allies could accomplish this at any time.[23] Hitler had decided a month earlier in favor of all-out production for the A-4 missile. As recently as January he had still been lukewarm about rockets, because even Himmler's repeated attempts to arrange the meeting with the führer—sought by Dornberger—went nowhere. Hitler had already been treated for the second time to one of Dornberger and von Braun's dog and pony shows, at his *Wolfsschanze* (Wolf's Lair) headquarters in East Prussia in August 1941, and apparently saw no need for another. The rocket program had not been able to stage a single successful flight of the A-4 until October 3, 1942, following two embarrassing (though normal for such complex new technology) failures, including one in June 1942 that nearly fell back upon Albert Speer.[24] Six months later, another crashed while Himmler looked on during his first trip to Peenemünde. Perhaps the blatant hazard of these tests was why Hitler never came to Peenemünde to watch one. The setbacks continued with a string of five more failures that did not abate until May 1943.[25]

A curious sidebar to the narrative of these ominous months of spring, 1943, is that von Braun chose the season to be wed, or at least tried to get married. Three documents survived the war showing that on March 25, two days after his thirty-first birthday, he identified himself as an *SS-Hauptsturmführer* (captain) on his "Dr. Wernher Frhr. v. Braun" Peenemünde stationery and asked the SS *Rasse und Siedlungshauptamt* (Race and Settlement) office in Berlin for marriage papers, signing his name under the typical closing "Heil Hitler!" He wrote that he wished to be married as soon as possible, because his fiancée's family had lost everything in a recent air attack on Berlin. He returned the completed application on April 5 with a personal note of gratitude to Himmler, who took a direct interest in maintaining racial purity by inspecting the ancestry of applicants for any Jews. In von Braun's handwriting (or that of an office worker with remarkably similar script) at the top of the letter appeared the adulatory, but not required, greeting: "Führer!" His chosen bride was twenty-six-year-old Dorothee Brill, a decidedly un-Junker name, born in the small town of Tübingen near Stuttgart and recently living in the upscale Berlin suburb of Steglitz, which until the war years was home to a sizeable Jewish population that in 1923–24 had included Franz Kafka. The marriage did not occur, for reasons that are unknown, and even the engagement remained unknown outside his closest coterie for the rest of his life. Certainly the work-obsessed von Braun would not have sought SS clearance to marry a woman he knew was tainted with Jewish blood, so it is reasonable to assume either that he did not know or that the marriage failed to happen for other reasons.[26]

On June 28, Himmler paid a second visit to Peenemünde.[27] An A-4 test was scheduled for the next day, so he sat up nearly all night conversing with Dornberger, von Braun, and several other senior members of the staff in the baronial Hearth Room of the officers' mess. According to Dornberger's novelistic account, von Braun (who, two months after his abortive SS marriage application, was surely wearing his SS uniform, as partially captured in a photograph the next day) led off by assuring the Reichsführer-SS of their total commitment. As the hours slipped by and each man talked about his specialty, they even mused a bit about space travel, if Dornberger is believed. Then, in response to a gutsy question from Dornberger—"Reichsführer, what are we really fighting for?"— Himmler launched into the full-blown racist-biological panoply of Nazi plans for violent subjugation of Europe. Dornberger kept asking tough questions and claimed that the answers sounded "monstrous" to the apolitical technologists. Whether or not this fantastic palaver ever took place as recounted by Dornberger, it offers unique evidence that von Braun was intimately exposed to the personal thinking of the Third Reich's second most fanatical leader. Afterward, neither he nor anyone else present that night decided that the time had come to curtail their involvement. Indeed, von Braun's SS rank rose from captain to major (*Sturmbannführer*) as of

June 28. The next day, they launched an A-4 for Himmler that veered crazily over the piney dunes and dived into the Luftwaffe airfield at Peenemünde-West, blasting a crater 100 feet wide and destroying three planes. For the second try, Dornberger escorted the Reichsführer by boat out to the safe remove of the Greifswalder Oie, from where they watched a successful flight.

It is no wonder that Dornberger and von Braun relied on a color movie when they finally won their audience with Hitler. They traveled to Wolfs-schanze again on July 8, 1943, when he was near the end of his rope in Russia. Hitler was so impressed by their presentation about the October triumph—and particularly by von Braun himself, who lectured "without a trace of timidity and with a boyish sounding enthusiasm"—that he bestowed top priority to production of the A-4, on a par with infantry tanks. Urged by Speer, he also awarded von Braun the coveted honorific title "Professor."[28] According to Speer's description of the occasion, Hitler suddenly turned "ecstatic," declaring: "The A-4 is a measure that can decide the war. And what encouragement to the home front when we attack the English with it! This is the decisive weapon of the war, and what is more it can be produced with relatively small resources." The movie must have been very good, indeed, or maybe the pills and injections Hitler took every day were causing side-effects.[29] When Hitler had first appeared after hours of delay, Dornberger had been "shocked" by how much he had physically changed since their last meeting, noting: "He looked a tired man. Only his eyes retained their life." Staring from an "unhealthily pallid" face, those eyes "seemed to be all pupils."

But after the thrilling film narrated by von Braun—it probably owed an artistic debt to *Frau im Mond*, with a sequence reminiscent of Fritz Lang's film showing the sliding, 100-foot-tall gates of the rocket assembly hall as a mobile frame rolled slowly out carrying a completely assembled A-4—Hitler "came to with a start." He proceeded to interrupt "impulsively" Dornberger's further explanations, asking whether the missile's explosive payload could be increased to 10 tons and monthly deliveries to 2000. Dornberger tried to tell him why this was impossible without more years of development, but "a strange, fanatical light flared up" in Hitler's eyes as he "obstinately bawled": "But what I want is annihilation—annihilating effect!" When Dornberger cautioned that the rocket designers had not thought of annihilation at the project's birth, Hitler "swung around in a rage and shouted": "You! No, *you* didn't think of it, I know. But *I* did!" Field Marshal General Wilhelm Keitel, one of the VIP audience that included Speer and General Alfred Jodl, then promptly changed the subject.[30]

Even if Dornberger's narrative was melodramatized for the sake of a book written years later, its details pointed to an unstable and perhaps psychotic führer. Sallow complexion, dilated pupils, and sudden mood swings suggested pharmacological effects. This was the leader who ordered that all obstacles be removed from A-4 mass production. It was certainly to Dornberger's advantage to cast blame at a manic führer for the ruinous pressure that engulfed his beloved rocket program, but there is too much evidence of Dornberger and von Braun's making fantastic promises over the years—the multimedia shows they staged were emblematic—to accept this dialectic. Besides, the rocket program already enjoyed as much bureaucratic priority as it could realistically use. It is reasonable to assume that what was now driving Hitler and other once-skeptical leaders into fanaticism about Wunderwaffen was the loss of hundreds of thousands of German soldiers in Russia. "The thing I had always feared was here," Speer wrote many years later with inimitable suavity. "Tangible success was imperiled by immoderate demands."[31]

Perhaps of more lasting interest than the bizarre movie date with Hitler was the fact that he stipulated that "*only Germans* should be employed" in A-4 manufacture.[32] His concern was secrecy, as well as retribution for using such a fearsome new weapon. "God help us if the enemy finds out about the business," Speer quoted him as saying.[33] But this decree did not last long. Though he assigned Speer to direct the rocket program at the end of July, Himmler convinced the führer in August that using concentration camp prisoners was a better idea. They were already in use, anyway. Himmler then stepped over Speer's head and appointed Brigadeführer-SS Hans Kammler, aged forty-two, the chief of SS construction who held a degree in civil engineering and had distinguished himself by building the gas-chamber extermination camps at Auschwitz-Birkenau, to spearhead the work. Kammler was a Polyphemus of the Third Reich, consuming lives wherever they came within reach. Dornberger compared him to "a *condottiere* of the period of the civil wars in northern Italy" and Speer called him "one of Himmler's most brutal and ruthless henchmen."[34] With this choice, the fate of Mittelbau-Dora's prisoners was sealed.

On August 28, 1943, the first 107 prisoners arrived at the Kohnstein ridge from Buchenwald.[35] Another 107 came on August 30, then 1212 on September 2, and then hundreds per day, almost every day, month after month. Railroad freight cars packed in fifty men each for the journey.[36] Their destination, an old anhydrite mine (for calcium sulfate, used to make gypsum and plaster of Paris) taken over by the government in 1935 to store oil and other strategic chemicals, was undeveloped for human habitation. With no barracks, privies, potable water, or sanitation of any kind, the prisoners were simply marched to the site and herded en masse into the tunnels for excavation. As an especially harsh Thuringian winter approached and their numbers mushroomed, they worked in the unheated, unventilated caves while blasting continued around the clock, slept when possible on bare gravel or damp straw, later in crude bunks stacked four-high, and defecated into oil drums cut in half with planks laid across. By the end of the year, 11,655 men had entered the Kohnstein in this fashion and at least 550 of them had perished, a figure that would rise into the thousands by springtime. The constant flow of new prisoners into Dora was necessary because the old ones died in droves under Dantesque conditions.[37]

Von Braun went to the site immediately, for three days between August 29 and September 2.[38] He was apparently the first Peenemünder to do so, indicating the importance of the supervisory role he would play there, which would not be obvious from his disconnected line in organizational charts.[39] A French survivor claims to have seen him inspecting the tunnels with an SS guard around the twenty-first of September.[40] He came back on October 8 and 9, which just happened to be days when very small numbers of prisoners arrived, for the first time not from Buchenwald: fifteen and twelve Italian military internees, respectively, from Stalag IX-C in nearby

Bad Sulza. This was no doubt a relatively palatable spectacle, if von Braun saw it at all, compared to the cattle cars from Buchenwald. His next trip was November 26, the day after prisoner shipments paused until the first of December. There were then about 10,000 men packed in the tunnels and it is unlikely that von Braun went anywhere near where they amassed, because catastrophic hygienic conditions and starvation were fueling rampant plagues of typhus, dysentery, and lung disease, with no medical care. Corpses were hauled back to Buchenwald for cremation until Dora got its own ovens in March 1944. From the beginning of January to mid-April 1944, 6000 incapacitated prisoners were also shipped out of Dora to the SS extermination camps at Lublin and Bergen-Belsen in Poland.

Albert Speer visited the tunnels—dubbed the *Mittelwerk*, or Central Works—with Kammler on a cold day in December 1943. His Ministry chronicle entry for that tour has often been quoted regarding the "barbarous" conditions he saw: "On the morning of December 10, the minister went to inspect a new plant in the Harz Mountains. Carrying out this tremendous mission drew on the leaders' last reserves of strength. Some of the men were so affected that they had to be forcibly sent off on vacations to restore their nerves."[41] When Speer arrived, a prisoner was about to be executed—an SS Terror measure conducted in front of the assembled workforce, accomplished by garroting the condemned with a stick in his mouth and hanging him—but he evidently prevented the atrocity on this occasion.[42] "The prisoners were undernourished and overtired; the air in the cave was cool, damp, and stale and stank of excrement. The lack of oxygen made me dizzy; I felt numb."[43] Yet, a week later, he wrote to Kammler lauding him for transforming the tunnels "from a raw state into a factory" in the "almost impossibly short period of two months," a feat that "does not have an even remotely similar example anywhere in Europe and is unsurpassable even by American standards."[44] Statements like this have caused historians to dismiss Speer as a wily liar.[45] A month later, Speer entered a clinic in a state of total exhaustion and collapse, but not before ensuring the construction of barracks outside the southern tunnel entrance, which were completed in the spring of 1944.

The compartmentalization of Speer's mind provides a possible template

for von Braun's thinking, but one that can be applied only with little documentary foundation.[46] There are no known pre-1945 statements about von
Braun's experiences at Mittelbau-Dora. His first comments after the war
came on October 14, 1947, when he gave a sworn statement for the defense
in the Dora war crimes trial of Georg Rickhey, Mittelwerk's general director from May 1944 to the war's end.[47] Von Braun testified that he had
never *worked* in the Mittelwerk factory, but had *visited* it fifteen to twenty
times between "September or October 1943," when the cave was still used
for oil storage, and February 1945 to discuss technical matters regarding
A-4 assembly. He said that working conditions in the tunnels were at first
"extremely primitive" and unsuitable for a production facility with several
thousand men. He said there was no camp for the prisoners and they
stayed in the primitive passageways. He then listed improvements that
were made by summer 1944, such as sanitation and ventilation, and said
that a prison camp was established outside the caves. The statement was
clearly aimed at demonstrating that Rickhey had presided over these so-
called improvements, but it revealed close knowledge of the entire situation for the full period of Mittelwerk's existence. That he chose to defend
a high official there suggests that the experience had not caused any crisis
of conscience, to say the least.

Von Braun's only other statement regarding Mittelbau-Dora—besides
several terse remarks in television and newspaper interviews or to hagiographic biographers in the 1960s and 1970s—came in another deposition
given on February 7, 1969, during the renewed trial in West Germany of
several Dora figures.[48] "I was never in the Dora prison camp," he testified
in German, though he visited the Mittelwerk "about 15 times." "In the
summer of 1943, when the blasting work for the [tunnel] extension had
already begun, but not production, I was in the underground galleries. At
that time, some of the prisoners were brought into these underground
tunnels. I walked through, with the visiting group, these temporary
underground lodgings. After this, quite precisely, I never saw these underground lodgings again. I imagine they were moved somewhere else after
they finished the Dora camp." He added: "During my visit in Mittelwerk
I never saw a dead man and never saw any kind of abuse or killing. Dur-

ing later visits, I heard rumors that some abuse had occurred and that some prisoners had been hanged in the underground galleries because of sabotage. I never saw this myself and don't know, either, who recounted this to me." Regarding sabotage, which was a prime issue for Mittelwerk managers and impossible for someone at von Braun's level to ignore: "I never received either a verbal or a written official announcement that any case of suspected or proven sabotage had ever occurred." Once again, at minimum, the comments implied extensive long-term exposure to the Mittelbau-Dora site, as well as the absence of any effort by von Braun to investigate the existence of atrocities.

For counterevidence to these statements, historians must turn to the recollections of surviving prisoners, a few of whom wrote memoirs of their experience that vividly portrayed the hellish conditions and attendant atrocities.[49] In the 1990s, a small number recorded specific allegations about von Braun in the form of notarized affidavits.[50] For example, Guy Morand, a French resistance fighter arrested in Bordeaux by the gestapo on August 3, 1943, and transferred from Buchenwald to Dora on March 15, 1944, testified in 1995 about an incident during the second half of 1944 as he worked on the rocket assembly line:

When I arrived to work with my shift, I noticed that my chronometer had disappeared, but it was found a little while later on the ground under another console. It was probably a stupid form of sabotage (because we in fact carried out much more elaborate forms of sabotage) by one of the deportees on night shift. This form of sabotage was all the more serious as it could in fact be considered as punishable in any case by immediate hanging. For that reason, I didn't want to get this comrade into trouble and so I tried to claim to the Meister [German civilian supervisor] that it was a stupid accident. Like the good Nazi he was, he immediately started shouting that it was sabotage, when just at that point von Braun arrived accompanied by his usual group of people. Without even listening to my explanations, he ordered the Meister to have me given 25 strokes in his presence by an SS who was there. Then, judging that the strokes weren't sufficiently hard, he ordered that I be flogged more vigorously, and this order was then diligently carried out, which caused much hilarity in the group, and following this flogging, von Braun made me translate that I deserved much more, that in fact I deserved to be hanged, which would certainly be the fate of the "Mensch" (good-for-nothing) that I was. Following this awful beating, of course, I couldn't sit down for over a month as we had little opportunity to get any treatment.

I think it is pertinent to point out that I knew who von Braun was because he had already come into our "hall" for rapid inspections, each time accompanied by civilians, no doubt engineers, high-ranking officers of the Luftwaffe or of the SS, all high-ranking officers, who followed him respectfully. The Meisters, who were themselves impressed, did not have to be asked twice to say that this was von Braun, one of the inventors of the "V2." During these unannounced inspections in the Dora underground factory, von Braun couldn't fail to see the abominable conditions of the deportees, and his inhumanity, I would say his cruelty, of which I was personally a victim, are, I would say, an eloquent testimony to his Nazi fanaticism.[51]

Likewise, Robert Cazabonne, another French resistance fighter arrested by the gestapo on August 16, 1943, who arrived at Dora as prisoner 21124 from Buchenwald in October of that year, wrote in a sworn affidavit in 1997:

In 1944, I was assigned to a Kommando [job group] that had the task of maintaining the air vents and pipes. We were therefore often called upon to go into the tunnel.

One day (I cannot be precise about the date), we were suddenly ordered to stop working, put along with other deportees and set to one side. There were rumors of sabotage among us. It was then that we saw several prisoners arriving in chains who were hanged from hoists. Some distance away there was a group of Germans (SS and civilians). One deportee, who was beside me, pointed out one of then to me, saying "that's von Braun."

As I didn't know this person, I can only say what I heard at the time.[52]

Another former prisoner, Georges Jouanin, reported being slapped by von Braun in 1944 for stepping on a fragile component while working inside a missile's tail section.[53] While these statements resound with authenticity, they suffer according to academic standards of historical evidence by their remove in time from the original event, lack of corroboration, and the possibility of mistaken identity. Such standards tend to work against the dispossessed, of course, in favor of those with the power and money to create archives. The *Häftlinge* were permanently handicapped in recording their experiences by the very state of being prisoners. Yet von Braun's sworn statements are just as problematic. There is no doubt that he was directly involved with the exploitation of slave labor, as shown by a letter to Mittelwerk chief planner Albin Sawatzki in August 1944, in which he discussed the transfer of certain skilled workers from Buchenwald.[54] The underlying truth, finally, is that an atrocity called Mittelbau-Dora existed and that von Braun had extensive first-hand experience of it as an overlord. Historians must ultimately decide whether to believe a man who spent the rest of his life keeping Dora in the shadows or those who spent theirs trying to shed light on it.

At the cost of thousands of lives, the first V-2 rockets rolled off the underground assembly line on New Year's Eve, 1943.[55] Speer was right to regard this accomplishment with awe, though Mittelwerk production chief Arthur Rudolph recalled many years later that they were declared *Pfusch* (rejects) and sent back underground for rework. Von Braun would spend 1944 devoted wholly to shepherding as many useable rockets as possible out of the Syracusean caves. Along the way he would join many other German armaments executives who were arrested capriciously by Himmler to remind them of who was really in charge, thereby gaining the lifelong excuse of self-preservation for having been a loyal Nazi weapon-maker. And when the end finally came, he would walk into the welcoming arms of the U.S. Army as though nothing evil had ever happened.

10

"A PSYCHOLOGICAL BLOCK"

O N THE MORNING of July 16, 1969, a rocket 363 feet tall and 33 feet in diameter, weighing 6,699,000 pounds with a payload of 1300 tons, rose over the sea from the sandy east coast of central Florida carrying three men on the first trip to the moon. The United States government had spent about $50 billion to aquire the rocket over the previous eight years, out of a total Project Apollo cost of 150 billion.[1] An American named Wernher von Braun, fifty-seven, from Huntsville, Alabama, was heralded as its mastermind.

Von Braun was a "rocket scientist" born and raised in Germany who had built that country's V-2 missiles during World War II and then expatriated to the United States to continue his work. He was a handsome, congenial personality on television and in glossy magazines, an inspirational figure who personified the "Space Age." This was the extent of what most Americans knew about him.

Going to the moon impressed millions of people around the world that American capitalism had beaten Soviet communism, which was the space program's original motivation.[2] But at home it was not the completely glorious event foreseen on May 25, 1961, when President Kennedy told Congress that the goal was part of "the battle that is going on around the world between freedom and tyranny." The decade had turned ugly, with assassinations, race riots, and a corrupt war in Southeast Asia. Multitudes of postwar youth who had been thrilled by triumphant feats in space during their childhood were alienated from American life by 1969. They did not subscribe to the freedom vs. tyranny template and were not inclined to separate technological achievements from politics.

As for Wernher von Braun, whom many remembered from his appearances on Walt Disney's science-fictional "Man in Space" television shows during the 1950s, he had already been satirized by a popular movie and a clever song that cast him as considerably less than noble.[3] These were criticisms from a counterculture, however. Though the young generation was large, so was the number of American veterans of World War II, who might have been expected to take offense at the rise of a former enemy weapons designer to celebrity status. Yet in the mainstream, as reflected by the iconic worldview of Time Inc. and network TV and official Washington, von Braun remained a pioneering genius, a baron of the stars who empowered many admirers to imagine reaching them someday. He died on June 16, 1977, while this was still a cultural enthusiasm.

It is fair to say that by the beginning of the twenty-first century, anyone much under the age of forty probably did not know who Wernher von Braun was. Like the Space Age itself, with its astronauts who had the right stuff and taxpayers willing to spend fortunes to send them off the earth as surrogates, he had faded into a kind of camp antiquity. His value in the aftermath of that era is therefore symbolic, as perhaps it always was, of the stupefying social power that technology can acquire and lose only to acquire again somewhere else. If it is difficult to reconjure now the 1960s imperative of walking on the moon, it is also mystifying how von Braun could have been a serial hero in two societies that had recently fought each other to the death. Scientists and engineers, in a kind of theological construct, evidently lay outside the sphere to which moral judgments applied.

VON BRAUN'S PATH to Huntsville began on the morning of May 2, 1945, when his younger brother, Magnus—whom he had hired as an assistant to keep him from combat and then sent into the Mittelwerk tunnels in September

1944 to supervise guidance mechanism work—coasted a bicycle down a
Tyrolean mountain road from Haus Ingeburg, a fashionable ski hotel at
Oberjoch on the German-Austrian border.[4] Magnus got off his bike, which
had a white handkerchief tied to its handlebars, and pushed it toward an
American soldier standing with rifle raised. Private First Class Frederick P.
Schneikert of the U.S. Army's 44th Infantry Division, a regimental interpreter from Sheboygan, Wisconsin, was on outpost duty with an antitank
company near the tiny village of Schattwald as German civilians and military personnel straggled through mountain passes to surrender. "We want
to see Ike," Schneikert remembered Magnus saying to him in English, refer-
ring to the Supreme Allied Commander, General Dwight D. Eisenhower.

The boyish twenty-five-year-old Magnus explained that "we" included
his brother Wernher von Braun, Walter Dornberger, and other inventors
of the V-2 rocket, who were waiting back at the hotel. "I think you're
nuts," Schneikert replied, "but we'll investigate." Eisenhower had diverted
American forces away from Berlin—to the consternation of Winston
Churchill—toward the Alps in search of an enigmatic "National Redoubt,"
a rumored mountain fortress of Nazi leaders and secret weapons. Schneikert had thus been alerted about false surrender tactics that led to
ambushes by SS and other die-hard guerrilla bands known to the GIs as
"werewolves."[5]

Magnus was taken under guard to the 44th Infantry's Counter Intelligence Corps post in Reutte, where a CIC officer who was aware that rocket
technicians were being sought for interrogation gave Magnus safe conduct
passes and instructed him how they should give themselves up. The
"National Redoubt" was turning out to be a mirage, but scientific intelligence teams—who knew, for example, that the Peenemünde wind tunnel
research station had been moved to Kochel, 25 miles south of Munich—
were intent on collecting whatever sort of human booty came their way. It
was a confused, uncoordinated process in which the military services
competed with each other for the spoils of war.

Schneikert recalled that when a convoy of the Germans finally came
down the mountain, they were carrying loads of wine and liquor hoarded
by the owner of Haus Ingeburg. During one of the parties that followed,

according to Schneikert, von Braun told him that America was going to
have a new national anthem now that President Roosevelt was dead.
"Then he went over to the piano and played 'The Missouri Waltz,' "
Schneikert remembered, a tune presumably in honor of Missouri native
Harry Truman, the new president since FDR's death on April 12. Schneikert then told von Braun that Germany also was going to have a new
anthem after the death of Adolph Hitler, which had been announced on
May 1. "I could play the piano a little then, too," Schneikert recalled, "and
I played *Du, du liegst mihr in Herzen* ('You, you lie in my heart')," a popu-
lar waltz at German-American weddings.[6] No Wisconsin nuptial ever
brought together two more unlikely cohorts.

This was Schneikert's gay memory of the event, at least. Besides the
verifiable date, place, and names, the atmospheric details were painted
long afterward with a brush colored by celebrity and cold war politics. If
there were in fact cases of wine and spirits, the GIs were no doubt happy
to consume them, but inviting freshly captured Germans to the party
defies credibility no matter who they claimed to be. If it happened, then an
extraordinary psychological situation was unfolding. The surrender of the
V-2 men marked not only the demise of the German Army's rocket program, but the launch of Wernher von Braun's American persona. It would
prove to be a remarkably impervious construct of convivial innocence for
the rest of his life, supported by the ignorance, wishful thinking, and
indifference of his audience, rather like the mass consciousness von Braun
had just left behind. That the rocket group had hurried away from a bes-
tial evacuation of Mittelbau-Dora's prisoners in which thousands perished
would not be detailed publicly for decades.

Well known already, of course, because there was no way to obscure it,
was that between September 8, 1944, and March 27, 1945, more than five
hundred V-2s had struck London, killing over 2700 civilians.[7] Yet this did
not affect the treatment of the "German scientists," as they were quickly
labeled.[8] Von Braun told a writer for *The New Yorker* magazine in 1951
that he never thought he would be arrested and punished. "We wouldn't
have treated your atomic scientists as war criminals," he was quoted as
saying, not opining who "we" might have been, "and I didn't expect to be

treated as one."[9] At the very least, the arrogance of his analogy matched that of Magnus's initial demand to "see Ike." The lack of American interest in whether he had been associated with slave labor during the war mirrored his own attitude about the subject. Unlike Albert Speer, he never developed even a specious, belated guilt complex.

On the morning after their surrender, the Germans were photographed like prized horses corraled by tired-looking American soldiers. Despite the mud of a raw alpine spring day, von Braun posed in clean leather coat, white shirt and tie, hat removed to display neatly combed hair and handsome face. Another photo showed him and Magnus together, smiling as though on a successful mountain holiday. They had obviously been well cared for at Haus Ingeburg, "living royally" as Wernher recalled years later, despite food shortages throughout Germany in the winter of 1944–45.[10] Because of a car accident in March while he was managing last-ditch technical efforts in the Mittelbau region, his entire left shoulder and arm were encased in a white plaster cast rising from his chest on a strut, resembling a fractured Nazi salute. But the clear message of the photographs was that the Germans were happy, healthy, unthreatening, and already kind of American.

After the photo session, they were transported to another Bavarian resort, Garmisch-Partenkirchen, with its famous view of Germany's highest peak, the Zugspitze, renowned as the locale of the 1936 Winter Olympics. At the U.S. Army's interrogation center there, formal questioning began of several hundred rocket scientists who had moved south from Peenemünde under Kammler's orders at the end of January as the Soviet Army advanced. Maintaining Peenemünde's hierarchy despite their status as internees, Dornberger and von Braun tried to keep the contingent—and their valuable expertise—unified as a bargaining chip. About half were released after preliminary interviews and failure to receive von Braun's imprimatur as sufficiently important to stay.[11] No one revealed the hiding place for the rocket program's technical archives. This occurred several weeks later during separate questioning of several Peenemünde specialists still in the vicinity of Nordhausen, when an astute interrogator led them to believe that Dornberger had approved releasing the location.[12]

The men had quietly discussed, while at Peenemünde late in the war when *Endsieg* (final victory) was no longer credible, the advantages of surrendering if possible to the Americans, whom they assumed could best afford to continue developing big rockets. The Americans would also have less to forgive than the British, French, or Soviets.[13] In this regard, nearly all were terrified of capture by the Red Army, as were millions of their panic-stricken countrymen as the advancing Soviets returned in kind the brutality that had been inflicted by German forces during Operation Barbarossa. Which Allied nation could spend the most money on rocketry in the future would have been an exceedingly esoteric consideration under the circumstances. "The Russians were only a hundred miles away, and we could already see that an Iron Curtain was coming down," von Braun told *The New Yorker* in 1951, appropriating Churchill's cold war metaphor that had nothing to do with the Soviet campaign to grind down the German army. "General Dornberger and I wanted our outfit to fall into American hands."[14] But it was probably as much fortuitous as intentional that their preferred scenario turned out to be. Kammler held greater authority than anyone else in the final twisted fabric of command over the rocket program, and when he ordered Dornberger to take about 450 key personnel to Oberammergau in the lower Bavarian Alps at the beginning of April, accompanied by *Sicherheitsdienst* (SD) guards, they obeyed, while suspecting he might try to use them as hostages in armistice negotiations.[15] If they had stayed in Thuringia, which the American Army swept across in mid-April, they could have met their favored captors a month earlier.

Of course, then they would have been connected by direct physical proximity to the hell discovered at Mittelbau-Dora and its subcamps around Nordhausen, where the war's final months produced a heinous slaughter. In March alone, the SS staged mass hangings of 162 laborers suspected of sabotage, resistance, or Communist sympathies, especially Russians. Repeating a savage pattern already established in the chaotic evacuation of death camps in Poland, 25,000 to 30,000 prisoners were shipped out of the Mittelbau area to hideously overcrowded Bergen-Belsen and other remaining concentration camps after Allied bombings of Nordhausen at the beginning of April heralded the inevitable end. When the

3rd U.S. Armored liberated Dora on April 11, they found only 600 skeletal survivors too weak from starvation and disease to move. Thousands of bony corpses were stacked everywhere and deep ash pits lay beside the smoldering crematorium. In a final massacre on April 13, SS guards still trying to destroy human traces of the world's first rocket factory locked 1046 exhausted prisoners in a barn in the nearby village of Gardelegen and burned them alive.[16] By getting out of the Breugelian Harz to the bucolic Alps, the Dornberger–von Braun "outfit" managed to reinforce the impression that they had no connection to these atrocities. The colleagues they left behind hunkered down to face whatever came.

Though he could not have known exactly what was transpiring to the north, other than that the Allies would inevitably come upon Dora and the tunnels, von Braun wasted no time in making his well-honed pitch about the future of rocketry. In an essay in German titled "Survey of Development of Liquid Rockets in Germany and Their Future Prospects"—with a byline using the rank of "Professor" that Hitler had given him—he wrote a brief bowdlerized history starting with the Raketenflugplatz, creating what would be a durable myth of nonmilitary genesis in which Dornberger's ordnance department was simply "interested in carrying on the work" of the cash-strapped spaceflight enthusiasts of Berlin. This would form the rubric of virtually every narrative written about him in the United States for the next three decades. In the second paragraph, he stated that "a complete mastery of the art is only possible if large sums of money are expended," the most accurate of the paper's predictions. "In view of the successful results achieved with liquid rockets, it was decided in 1936 to begin with the construction of a large experimental establishment for rocket development at Peenemünde," he continued without mentioning that the only workable devices at that time were "Max" and "Moritz," that Peenemünde was meant from the start to be a secret base for building advanced weapons for Hitler's use in a European war. After short technical commentary about the A-4 (never calling it the V-2), which highlighted the constant attempts to improve accuracy, he set free his imagination.

First, using present-tense verbs, he described the A-9, a follow-on to the A-4. The A-9 "has wings, which enables it to glide through the stratosphere," he claimed. "This enables the flight path to be increased to such an extent that the range of the A-9 is nearly double that of the A-4, that is, approximately 600 km." A small group of Peenemünde engineers had in fact studied the possibility of a winged A-4, especially after the Allies' D-Day invasion in June 1944 jeopardized V-2 launch sites in northern France that were within range of London. They had concluded early on that it would not work, primarily due to insurmountable aerodynamic control problems. But like every aspect of A-4 development, the A-9 program was distorted by pressure from a deteriorating military situation. Desperate shortcuts, such as simply welding swept wings onto an A-4 body and tail, resulted in one test rocket that crashed shortly after liftoff and another whose wings sheared off downrange. A prominent Peenemünde engineer later recalled the A-9 as an "eye-wash" project intended to convince impatient authorities that useful work was being done.[17]

If the A-9 was eye-wash, there was perhaps no polite term for what followed. "As a further development, it was intended to design the A-9 winged rocket to carry a crew," von Braun wrote. "For that purpose the rocket was to be equipped with a retractable undercarriage, a pressurized cabin for the pilot, manually operated steering gear for use when landing, and special aerodynamic aids to landing." This was Jules Verne territory, as was the A-10, a two-staged rocket with an intercontinental range of 5000 km "both in the piloted and pilotless versions."[18] During drawing-board studies at Peenemünde about increasing the A-4's range of about 150 miles, theoretical trajectories for a two-staged rocket that could reach New York had been calculated, but the few Projektenabteilung (Projects Office) engineers who thought about it understood how futuristic it was. "There were a lot of brain bubbles around," one remembered many years later. "And I tend to see the A-10 even as a bit of a brain bubble."[19]

Von Braun was not finished with his brain bubbles, however. "In the more distant future," he offered "long-range commercial planes and long-range bombers" that would cross the Atlantic in forty minutes; "multi-stage piloted rockets" that would go into orbit around the earth "with very powerful telescopes"; stations in "interstellar spaces" whose "erection should be easy" by men "who would float in space wearing divers suits" and which, "according to a proposal by the German scientist, Professor Oberth," could carry a mirror with "a diameter of many kilometers" for concentrating the sun's rays to change terrestrial weather or "generate deadly degrees of heat at certain spots." Finally, "it will be possible to go to other planets, first of all to the moon." In an especially provocative sentence, given that Hiroshima was still three months away, he then wrote that "we see possibilities in the combination of the work done all over the world in connection with the harnessing of atomic energy together with the development of rockets, the consequences of which cannot yet be fully predicted."[20]

It is obvious in retrospect that there was something about von Braun that encouraged other men—from Hitler to the young American intelligence officers at Garmisch-Partenkirchen—to take his fantasies seriously. His "future prospects" had been part of pulp science fiction for generations, but he was not dismissed as a fantasist. To the contrary, the Americans accepted him as a real professor with a PhD, a confident aristocrat, most of all as a practical armaments engineer. Rudolph Nebel and Walter Dornberger had capitalized upon his potent powers of suggestion very early on, and now it was the Americans' turn. The V-2 was such an astonishing leap of invention that it made von Braun's incredible forecasts seem within reach. What he was already injecting into the postwar years, starting at Garmisch-Partenkirchen, was not just unique technological expertise from his V-2 experience, but the infectious chimera of space travel from his childhood. The former was of immediate value to the military, the latter a thrust of imagination that might help spur technological progress, but also push it into romantic ventures and ultimately dissipate, like all dreams. To von Braun's immense good fortune, it did not dissipate for another thirty years.

Because no evidence of von Braun's private thoughts survived the war, and his later writing had the quality of public relations, it is difficult to know for certain whether he used space travel dreams cynically or ingenuously. If he had written his essay before surrendering to the U.S. Army—as a personal journal entry, say, during the war or the heady 1930s—it might have served as the kind of concrete evidence of visionary purpose that is missing from his pre–May 1945 archive (excepting schoolboy science-fictional stories), though often attested to by members of his circle. After 1945, space travel dreams served to sugarcoat what he had really been doing between the ages of twenty-one and thirty-three: developing weapons for the Nazi regime. The essay also perhaps told the story of his life in Germany as he wished it had been, a narrative cleansed of evil and defeat. Millions of Germans would have welcomed the opportunity to recast their lives in this fashion, but he was handed it on a platter. "The amusing thing about my country's collapse was that the V-2 crowd had its choice of what to do," he told *The New Yorker* with what the magazine's reporter described as a "jolly laugh."[21]

It is also clear that von Braun and his German colleagues were exhibiting what Hitler biographer Ian Kershaw has called "a psychological block on recognizing responsibility for their actions."[22] The complete disintegration of the Nazi value-system, which they had adhered to if not fully supported, rendered them mute in terms of expressing remorse. Kershaw found that few Germans who were forced to account for their conduct under Hitler, whether they were heavily incriminated leaders or from "ordinary" walks of life, could bring themselves to acknowledge their personal contributions to the Third Reich. This phenomenon was noticed by the interrogators at Garmisch-Partenkirchen. "There is almost nowhere any realization that there was something basically wrong with Germany's war or the employment of V-weapons," wrote intelligence officer Walter Jessel in a June 12, 1945, report. "There is recognition of Germany's defeat, but none whatsoever of Germany's guilt and responsibility."[23] Jessel had fled Germany in the 1930s and immigrated to the United States, becoming an American citizen and army second lieutenant. He was thus in a position to make such observations, and his reports on the rocket personnel stand out in retrospect for their skepticism. (He also warned of a conspiracy involving von Braun and Dornberger to withhold information, advising that it would be an "obvious absurdity" to give the rocket group security clearances.) For von Braun and many of the Peenemünders, muteness lasted a lifetime. The word "regret" was attributed to von Braun in the 1951 *New Yorker* interview, but it was couched in a preposterous claim:

"The Allies had bombed us several times at Peenemünde, but we felt a genuine regret that our missile, born of idealism, like the airplane, had joined in the business of killing. We had designed it to blaze the trail to other planets, not to destroy our own."[24]

While he was crystal-ball gazing in Garmisch-Partenkirchen, the first priority of the U.S. military was to win a war still very much alive in the Pacific. On May 26, 1945, Undersecretary of War Robert Patterson wrote a memorandum for Secretary Henry Stimson and the General Staff on the subject of "German Scientists."[25] "I strongly favor doing everything possible to utilize fully in the prosecution of the war against Japan all information that can be obtained from Germany or any other source," he began. He then listed several concerns:

> These men are enemies and it must be assumed that they are capable of sabotaging our war effort. Bringing them to this country raises delicate questions, including the possible strong resentment of the American public, who might misunderstand the purpose of bringing them here and the treatment accorded them. Taking such a step without consultation with our Allies, including the Russians, might lead to complications.

In addition:

> It is also my feeling that the information from these men should be obtained to the maximum possible within Germany and only those should be brought here whose particular work requires their presence here. It is assumed that such men will be under strict surveillance while here and that they will be returned to Germany as soon as possible.

At the moment that this and other policy ruminations on the matter were circulating around the War Department's upper echelons in Washington, the U.S. Army was stripping V-2 equipment as fast as possible out of the Mittelwerk "magician's cave," as one American intelligence officer on the scene called it, employing even former Dora prisoners strong enough to help.[26] Since the V-2 threat had become obvious in 1944, the army had started its own rocket program, known as Project Hermes, and was intent on launching salvaged German missiles within a vast wilderness at White Sands, New Mexico. For this they would need not just V-2 components, but the "German scientists" who knew how to handle them. All "delicate questions" about the Germans, as well as consultations with Allies, would quickly fall aside in the rush to haul away the rocket booty.[27]

One of the most delicate questions that von Braun had to address promptly was his SS membership. On Walter Jessel's June 10 "Qualification Sheet for German Scientific Personnel" for von Braun, under the heading "Any implications in War Crimes or other Nazi activities," he wrote: "Entered [Allgemeine (typed above caret mark)] SS (under pressure, he claims) in 1941. Last rank: Stubaf (equiv Major)."[28] This information must have come from von Braun himself, because there was no access yet to German government records in Berlin. Under the heading "Remarks," Jessel added: "Attitude characterized by following quote: 'I always was a German and I still am.' Considers Germany dead as a nation. Only [German (penciled above caret mark)] hope: To cooperate with western Allies to act as bulwark against eastern hordes, and as a beachhead for US and British forces in the coming struggle." Besides regurgitating the Nazi's racist label for the Soviets, von Braun apparently revealed in short order the mixture of nationalism and opportunism that had propelled him through the war, as well as the anti-Communism that would serve him well in years to come.

Though he appears to have lied about when he joined the SS, which was actually 1940, he was not foolish enough to hide the fact. What he meant by "under pressure" would not be explained on the record until his sworn "Affidavit of Membership in NSDAP" two years later. In any case, there seems to have been no deeper scrutiny of his SS affiliation. Not until after he arrived in the United States in September would a military intelligence report note, under "Record of Arrests": "1944 arrested for 14 days by Gestapo (Secret Police), Stettin, because of alleged remarks that B.

[Braun] had no interest in the rocket as a weapon."[29] This was the seed of a story that would expand to shield him for the rest of his life from accusations of having been an ardent supporter of the Hitler regime. That virtually all of the story's details derived from his or Dornberger's post hoc writing never seemed to matter.

The core of the story, substantiated to the extent that Albert Speer's memoirs can be believed and tangentially by a short cryptic note written by General Alfred Jodl, was that von Braun had been arrested in March 1944 by Himmler's agents and imprisoned for more than a week. Because there are no known contemporaneous records of this action, historians have been forced to make educated guesses about when exactly it occurred, who was involved, and indeed whether it really happened at all. Speer maintained in *Slave State* that Himmler habitually intimidated leading armaments makers with threats of arrest or actual incarceration as the SS munitions empire expanded and the Reich's fortunes on the battlefield plummeted, in a fashion that often ignored the practical effect upon production output of such bullying.[30] Speer wrote that during their rivalry to control weapons production, in an attempt "to show me that he could strike as he pleased even against important groups of individuals," Himmler "had Wernher von Braun and two of his assistants arrested" on March 14. The official justification, according to Speer, was that "these men had violated one of my regulations by giving peacetime projects precedence over their war-production tasks." Von Braun and his staff "used to talk freely about their speculations, describing how in the distant future a rocket could be developed and used for mail service between the United States and Europe," Speer explained. A brain bubble, in other words, may have opened the door for ruthless Himmler, who wished to crack his whip over the rocket program and, by extension, Speer.[31]

Years later, von Braun and Dornberger each wrote novelistic accounts of this episode, complete with extensive fabricated dialogue.[32] Von Braun claimed that a trepidatious meeting with Himmler in February 1944, during which he spurned the Reichsführer-SS's invitation to help lift the rocket program out of army "red tape," precipitated the arrest, along with allegations that "I had space-travel in mind when [the A-4] was developed."[33] Dornberger embellished the space-travel angle to include himself, and took credit for springing von Braun and the others from prison. (Speer maintained more convincingly that he had secured their release directly from Hitler, when the führer visited him on March 18 at Klessheim Castle, Hitler's guest house near Salzburg, where he was recuperating from a lung embolism. A week passed before they were actually freed. This suggested that Himmler, in typical fashion, had seized the opportunity of Speer's prolonged illness to make a move on the rocket program's top management, using trumped-up charges of treasonous utterances as an excuse. The SS doctor assigned to Speer, Himmler's personal physician Karl Gebhardt, was circulating false reports to Hitler and Göring that Speer was terminally ill. To interfere with a program as prominent as the A-4, it was important for Himmler to get Speer out of the way, because an ordinance issued by Hitler in March 1942 had stipulated that "any legal proceeding for damage to armaments could take place *only* at my instructions," as Speer noted about SD-generated arrests. From the early spring of 1944 onward, because of the unmistakeable collapse of German forces on the eastern front, the reach of the SS-SD-gestapo net became more and more desperate.)[34]

As for the Jodl note, which he jotted during an intelligence briefing on March 8, it indicated that Sicherheitsdienst spies at Peenemünde had reported remarks "that the war will turn out badly" and "the main task is to create a spaceship."[35] Dornberger claimed that in Zinnowitz, where parties were often held for the rocket workers, these secret agents "by taking words out of their context had twisted them to appear treasonable."[36] By this point, of course, it was obvious to everyone besides utter fanatics that the war was going badly. The men's tongues had apparently been loosened by wine at the elegant old Schwabes Hotel. Speer explained that the feared SD "maintained a widespread undercover system, which had its agents in almost every office and factory," but in order to justify itself bureaucratically the network "developed a kind of compulsion to keep presenting new complaints," which was "bound to lead to exaggerations." It was "very tempting to use such an instrument for political purposes," he observed. "Especially in the case of armaments."[37]

Dornberger's memory and Jodl's note matched in naming Klaus Riedel (who helped develop A-4 deployment equipment), and Helmut Gröttrup (assistant to guidance, control, and telemetry director Ernst Steinhoff) as the two men arrested with von Braun. Sometime after the war, von Braun added his brother, Magnus, to the jailed group.[38] Riedel and Grötrupp had been solid contributors at Peenemünde, but hardly priceless to the rocket program ("Grötrupp I don't know so well," Dornberger claimed to have told Field Marshal Wilhelm Keitel while trying to free them). Riedel was the former Raketenflugplatz machinist who had initially opposed joining Dornberger's military effort in 1932, then worked for Siemens before finally going to Peenemünde in 1937 under von Braun's umbrella, where he helped create V-2 mobile launchers after von Braun pulled him off test stand work for being incompetent. He died in a suspicious auto-mobile accident near Karlshagen on August 4, 1944.[39] Gröttrup stayed behind in Thuringia during the war's final weeks and eventually figured in the Soviet Union's postwar rocket program.[40] Both men had vague rep-utations as leftists, though this would not have taken much in their milieu. Jodl's note linked Riedel to the *Liga für Menschenrechte* (Human Rights League, banned by the Nazis in 1933) and Gröttrup to the *Pan-Europa-Bewegung* (Paneuropean Movement, likewise banned), which was enough for Jodl to call them an *edelkommunistisches Nest* (refined Communist cell). In any case, all of the charges were buffoonish and a transparent excuse for Himmler's incursion.

For the fortunate von Braun, however, the arrest turned out to be one of the luckiest strokes that ever befell him. It featured in myriad U.S. gov-ernment security reports for many years after the war, cited as proof of his overarching belief in space travel, merely opportunistic career building weapons for Hitler, and general antipathy toward the Nazi regime. Nei-ther American intelligence investigators, whether employed by the army or FBI, nor the countless journalists and public-relations biographers who wrote about him for the rest of his life, were inclined to shed light on the story's factual basis or historical context.

After the initial interrogation of von Braun was finished at Garmisch-Partenkirchen in the middle of June, the army flew him back north to help collect more rocket personnel and their families before the Soviets took control of the Nordhausen region (in keeping with the Allies' Yalta Agreement of February).[41] Many were moved in railroad cars to make-shift quarters in the town of Witzenhausen, just inside the American zone, some 40 miles southwest. There was little concern about the legality of the roundup or what would ultimately be done with the people, only that the territory be swept before the Russians arrived at the beginning of July. On July 20, the Joint Chiefs of Staff in Washington approved a secret program code-named "Operation Overcast" to import German scientists temporarily "to assist in shortening the Japanese war and to aid our post-war military research."[42] Known or alleged war criminals were specifi-cally excluded and any discovered subsequently were to be sent back to Europe for trial—a clause that would be skirted. The War Department made a short public announcement saying simply that some German sci-entists and engineers were being brought to the U.S., then clamped a tight lid on further information.[43]

The British had their own code-named operation that summer to exploit the rocket scientists, called "Backfire," aimed at launching salvaged V-2s over the North Sea from Cuxhaven at the mouth of the Elbe River. Dornberger, who not only lacked von Braun's charm but was roundly despised as the Wehrmacht general responsible for firing V-2s, was shipped first to Cuxhaven and then to London and eventual internment at the Island Farm prisoner-of-war camp in Bridgend, Wales, for the next two years. Overcast thus left behind the father of the German rocket program, who was not among the initial group of some 125 transferred to the United States. Magnus von Braun, who possessed only minor experience, was.

In August, Wernher went to London for interrogation, too, at the behest of Sir Alwyn Douglas Crow, who was in charge of developing rockets in the British Ministry of Supply.[44] "I must admit that I thought the British might be unfriendly to me," he was quoted as saying in the 1951 *New Yorker* article, "but I was hardly inside his office before we were engaged in friendly shoptalk." Billeted for two weeks at a detention center near Wimbledon, he was delivered to the Ministry each day by an RAF driver, who once stopped their car in front of a downtown building that had been torn apart by a V-2. "It looked as if it had been a six-story office building, but I was unable to tell the precise way in which the V-2 had done its damage, because the rubble had been cleared away," von Braun recalled. His only reaction to the carnage was to wonder what had become of the German undercover agents in London who radioed damage reports. "I never did find out, but one thing I know is that we had some good ones there. Our battery commanders on the French coast used to have reports on V-2 effectiveness within an hour after a rocket had been launched." Seven years after the war, his thoughts remained frigidly technical, at least as given to *The New Yorker.*

Back in Germany, on September 15, the Headquarters/United States Forces/European Theater issued an order "by Command of General Eisenhower" that "the civilians named below, German, will proceed on or about 18 Sept 1945, from their present station in this theater by first available air transportation to the United States" for "temporary duty."[45] Von Braun had finally received Ike's direct attention. The order's stipulation that "upon completion of this duty" the men "will return to their proper station in this theater" would be ignored. Besides von Braun, there were six other rocket men among the sixteen scientists and engineers on the list, including Eberhard Rees, a manufacturing and production chief who had stayed in Nordhausen. On the same day, von Braun signed a War Department employment contract paying him 31,200 marks (in Germany, not the United States) for a minimum of six months and a per diem of six dollars for the entire period.[46] On a copy dated September 14 whose entries were handwritten, he listed his home address as "Elektromechanische Werke Karlshagen," the misleading name given to the Peenemünde base in August 1944 when it was nominally privatized while the SS and Speer's Armaments Ministry were vying for control. The final typed contract showed the address in more detail as "Elektromechanische Werke GmbH, Karlshagen, Peenemünde, Haus 18." As dependents he named his father and mother, whose whereabouts were unknown to him, plus Klaus Riedel's widow and three-year-old daughter. The contract promised adequate housing and food in Germany for dependents and an effort to locate them if necessary.

On September 18 as commanded by Ike, the men flew on a military transport from Paris to Delaware and then to Boston on September 20, where Project Overcast had placed concealed administrative offices in the harbor's Civil War–era Fort Strong. Von Braun noticed the first symptoms there of what developed over the next few weeks into a serious case of hepatitis. From Boston the group was escorted by Major James P. Hamill—a twenty-six-year-old Fordham physics graduate who had supervised the seizure of V-2 components—on a train to Baltimore, where von Braun's colleagues were sent to the Army's Aberdeen Proving Ground north of the city on Chesapeake Bay. They set to work there processing the Peenemünde document treasure, only the first of many occasions when the Germans had an opportunity to decide what was important and what was not among the technically valuable but potentially incriminating papers (surviving documentary evidence of Mittelwerk sabotage reports, for example, which could have led to prisoner executions, is remarkably thin). Von Braun continued on with Hamill to Washington and finally to El Paso, Texas, and nearby Fort Bliss, in the first week of October.[47] A five-year stay in the desert lay ahead before finding a permanent home in the green hills of Alabama.

In just five months, he had moved from the snow-capped Alps to the scrub-covered Franklins in one of the most desolate stretches of North America, so remote that the Trinity A-bomb test had been held across the state line in New Mexico in July. Hamill's V-2 trove was dumped in the desert about 50 miles from Fort Bliss, waiting for its former master. So far from its origins, the gear must have looked like those strange souvenirs that puzzle travelers when they return from an exotic land. For the rest of October, von Braun could only contemplate the turn in his own life from a military hospital bed. He might well have thought that Germany no longer existed, which in a way was true. But he was a resilient young man and the future was his, more than ever.

EPILOGUE

COUNTLESS IMMIGRANTS HAVE come to America expecting to find the streets paved with gold, and the "German scientists" were no different. Instead of ensconcing them at a New World Peenemünde, however, the U.S. Army dropped them off in a crude cowpoke wasteland where the most notable recent events were the explosion of atomic bombs and the invention of the margarita cocktail.[1] They were not so much put to work there as stashed where no other country could get at them—particularly Great Britain, France, and the Soviet Union, which had their own shopping lists for German expertise. One of their disarmed Wunderwaffen was put on display along Pennsylvania Avenue in Washington, DC, under a billboard that read "This is a V-2 rocket seized by U.S. Army Ordnance," but its inventors were hidden far from civilization in a guarded camp.[2] The primary rationale for bringing them over in the first place had disappeared when Japan surrendered, but the reasons were multiplying all the time.[3]

Wernher von Braun's colleagues began to arrive from Germany at the end of 1945.[4] They traveled across the Atlantic in Spartan troopships, not one of Donald Douglas's transoceanic DC-4 "Super Mainliner" airplanes as had their boss (albeit in a military transport configuration). Their first job was to start what they had recently been forced to stop—constructing and launching V-2s, now using the boxcar-loads of jumbled parts for some 100 missiles that had been laid out in the Tularosa Basin's caustic desert environment.[5] Basically, they were to nurse the American military and civilian participants in Project Hermes along the learning curve that had consumed their attention since 1942. "That job took eight months," von Braun later recalled. "We seemed to be expected to do it in two weeks."[6] Only about half of the roughly 6000 V-2s produced in Germany were ever launched during the war, so the experience at White Sands in 1946 was somewhat frustrating for the military, industrial, and academic boffins who converged there to try out the famous rocket. The V-2 trove was rapidly turning into scrap metal. Eighty miles from El Paso, moreover, the living conditions were rather less civilized than at the former Nazi showcase on the Baltic seashore.

The first shot on April 16 flew out of control, shed a tail fin, and crashed at close range. The second and third reached higher altitudes, but smashed to smithereens in deep craters on impact (as designed), pulverizing the technical payloads placed aboard by eager scientists like James Van Allen from the University of Iowa and teams from the Johns Hopkins Applied Physics Laboratory. The V-2 was a battlefield munition with lots of problems, they soon appreciated, not a refined scientific instrument. By the end of 1946, the launch failure rate was more than 33 percent.[7] "Frankly, we were disappointed with what we found in this country during our first year or so," von Braun later said.[8] Still, it was not such bad duty at a time when German cities lay in ruins, many of his countrymen were destitute and starving, millions were nothing but ash, and former leaders awaited execution for war crimes. And it got steadily better.

Not all American scientists were keen to use the V-2 booty and the onetime enemies who came with it. In January 1947, for example, prominent faculty members at Cornell—such as the renowned German émigré Hans Bethe, who had directed the Manhattan Project's theoretical division, and illustrious aerodynamicist William R. Sears—spearheaded a protest against the War Department's importation of German scientists and engineers.[9] "The fact that these men were directly or indirectly linked with a regime whose infamous record included, among other things, the

most brutal persecution of free science must fill every citizen, and in particular every scientist, with deep apprehension," stated a resolution sent to
the Federation of American Scientists, advising that the Germans be sent
back to where they came from when their work was finished.[10] A month
earlier, following War Department–sanctioned publicity about Operation
Paperclip, Overcast's successor, forty luminaries—including Albert Einstein, A. Philip Randolph, Norman Vincent Peale, and Rabbi Stephen
Wise—had sent a telegram to President Truman, protesting that the Germans' "former eminence as Nazi Party members and supporters raises the
issue of their fitness to become American citizens or hold key positions in
American industrial, scientific, and educational institutions."[11] This was
exactly the kind of public reaction the military had sought to head off
with strict secrecy. In the summer of 1947, Rep. John D. Dingell, Democrat from Detroit, gave the protest a populist voice on the floor of the
House of Representatives, saying, "I have never thought that we were so
poor mentally in this country that we have to go and import those Nazi
killers to help us prepare for the defense of our country."[12] Perhaps to some
small extent, certain elements of the U.S. military agreed with this line of
thought, noting in a September 18, 1947, security report that Wernher von
Braun "is regarded as a potential security threat," though "not a war criminal" based on available records.[13]

But the protests were evanescent, indecisive, handicapped by the secrecy
of government policies, and swamped by larger issues. In addition to the
first rumblings of cold war rivalry with the Soviet Union, there were business imperatives in Washington, as always.[14] Leaders of American industry and their trade associations, many of whom had served in intelligence
units that had cherry-picked all over Germany for superior technology
and expertise, successfully lobbied President Truman for a commercial
exploitation program, which ran until it was shut down in 1947 for the
sake of German economic recovery.[15] Military budgets naturally plummeted in the immediate postwar years, leaving Fort Bliss as bleak as ever,
and the decimated aviation industry jumped at the chance to capitalize on
German technology. "Very early on we became involved with von Braun
and his associates when they were stationed at Fort Bliss (surely a euphe-

mism)," remembered J. Leland Atwood, president of North American Aviation, whose nascent Rocketdyne division in Los Angeles would become a premier producer of large rocket engines. "Our rocket work was, in large measure, built on the Peenemünde V-2 model to start with."[16]

Early in 1946, Wernher learned that his parents were alive, albeit completely dispossessed, in Silesian territory now part of Poland, thus reenacting the centuries-old ebb and flow of Junker fortunes. The family's experience with gaining and losing estates was truly prodigious. Wernher was allowed to return to Germany under round-the-clock military guard in order to marry his eighteen-year-old first cousin, Maria von Quistorp, in March 1947. On the same trip, he collected the baron and baroness, who immigrated along with his bride to El Paso that month under the dependent provisions of Operation Paperclip.[17]

Gradually, von Braun's army handlers loosened their grip. Top officials in Washington from the Oval Office down fell into line with the notion of Germans as "intellectual reparations," and life moved on with amnesiac velocity. In December 1947, the Dora war crimes trial at Dachau ended and its proceedings were classified, the army having helped von Braun to avoid in-person testimony. The Peenemünders were wise enough to shut their mouths and self-censor their own war stories, while their military milieu took care of placing their files under lock and key. The perceived value of their knowledge continued to trump any other considerations. The British released Walter Dornberger, who came to the United States under the U.S. Air Force's newly independent auspices (and by 1950 was an executive at Bell Aircraft in Buffalo, New York). As early as July 1947, *Popular Science* magazine boasted that the U.S. Navy's homegrown Viking missile would double the V-2's altitude record and weigh much less than "one of those Nazi dinosaurs."[18] Just four years after the last V-2 fell on London, the British Interplanetary Society named von Braun an Honorary Fellow in 1949. Technological progress itself was making the "Nazi scientists" seem both less magical and less dangerous. To the lifelong benefit of most of them, it was also making what happened during the war more unbelievable.

On September 23, 1949, President Truman announced that the Soviet Union had tested an atomic bomb. On October 1, Mao Zedong established the People's Republic of China. In November 1949, von Braun's extralegal immigration status was normalized with a proper visa by riding a trolley back and forth across the U.S.-Mexico border at Juarez.[19] On June 25, 1950, North Korean troops attacked across the 38th Parallel into South Korea. For the last few years, von Braun had so much time on his hands that he wrote a science-fiction novel of nearly 500 manuscript pages, but the world suddenly changed very much in favor of a longtime anti-Communist rocket builder. The novel, titled "Mars Project"—in which seventy passengers go to Mars in ten spaceships after the West defeats the East with atomic bombs dropped from an orbiting space station—was an amateurish brick of 1920s space-travel fantasies and 1930s Nazi propaganda about the Bolshevik menace. It is safe to say that the many New York editors who turned it down could not appreciate how well it expressed a dream of what an unspoiled Peenemünde might have accomplished had Hitler fought the ultimate war of annihilation against the Asian hordes. Von Braun would eventually sell the rights to a German publisher, which had it rewritten by a Luftwaffe veteran and illustrated with *Frau im Mond*-style pictures.[20]

In the spring of 1950, the Germans left Fort Bliss behind and transferred to Redstone Arsenal in Huntsville, Alabama, a former chemical munitions plant that Senator John Sparkman was godfathering out of its postwar doldrums. Von Braun moved with his wife and toddler daughter, Iris. Another daughter, Magrit, arrived in spring 1952 (a son, Peter, came in 1960). In addition to becoming a family man, he had found God. He told *The New Yorker* that "as long as national sovereignties exist, our only hope is to raise everybody's standards of ethics." "I go to church regularly now," he added. "Did you at Peenemünde?" the reporter asked. "I went occasionally," he replied. "But it's really too late to go to church after a war starts. One becomes very busy."[21]

In 1952, he began an enormously successful sideline as a popularizer with a series of lavishly illustrated articles—pre-screened by the Defense Department, like all of his outside writing—in *Collier's* magazine that let loose the same flights of imagination he had released at Garmisch-Partenkirchen, now amplified by the power of American advertising.[22] The vivid full-color pictures of manned spacecraft, lunar and planetary exploration, and giant wheel-shaped orbital stations, with breathless commentary, struck a nerve of pleasure in the American public similar to what a later generation would experience at *Star Wars* movies, propelling von Braun into the heavens of mass media promotion. The *Collier's* phenomenon led to similarly cathartic appearances in 1955 and 1957 on Walt Disney's popular television shows that promoted "Tomorrowland" at the Disneyland theme park. On November 30, 1956, he appeared on comedian Steve Allen's popular television show. Fan mail inundated his office, much of it answered in colloquial English by a public affairs man, though von Braun was a sponge for American slang. On March 13, 1958, he and Maria dined with Washington socialite Perle Merta, the famous "hostess with the mostess." Had nothing else ever happened to von Braun in America, the *Collier's* and Disney exposure would have cemented him in the minds of baby boomers in the way that Luke Skywalker took hold of mass consciousness a generation later. The illustrations and animated images were pseudoscientific, but they helped to sell an adventure to a gullible or skeptical audience. Von Braun had done it all before, of course.

By 1953, Redstone Arsenal was well on its way to being the American incarnation of Peenemünde, with von Braun installed as civilian director at the Army's Guided Missile Development Division of the Ordnance Missile Laboratories, under the command of a Brigadier General, Holger Toftoy—essentially the same organizational scheme and vocabulary that the Wehrmacht had used. Huntsville would soon be called "the German part of the state" by native Alabamans, as the Peenemünders transplanted their cultural proclivities for music and literature into the provincial Southern town.[23] They did not perturb the racially segregated society that the state epitomized and were as insulated from any center of liberal thought as they had been in El Paso. On April 14, 1955, von Braun became a U.S. citizen.

As at Peenemünde, the men developed weapons, not spaceships. Their first product was a liquid-fueled rocket dubbed the Redstone, essentially

an updated A-4 with a nuclear warhead. Redstone finally manifested
the massive destructive potential of guided missiles that the convention-
ally armed V-2 had only implied. It was followed by the more powerful
Jupiter—known as an IRBM, or intermediate range ballistic missile, as
compared to the ICBMs, or intercontinental ballistic missiles, being developed elsewhere—which devolved into a byzantine turf battle between
the army and air force over which service would have the biggest, longest
rocket and thus dominate the new field's gigantic budgets. During this
growing bureaucratic turmoil, which must have felt more than vaguely
familiar to the Peenemünders, both the United States and the Soviet
Union announced that they would try to launch a satellite around the
earth as part of the 1957–58 International Geophysical Year research program. The army was soon ready with von Braun's Jupiter-C—a modified
Redstone booster with two smaller stages on top—but the navy's far less
mature follow-on to Viking, called Vanguard, got the nod from Washington instead, much to his disappointment. As a result, the nation received
its biggest political shock since learning about the Russian A-bomb when
a satellite called *Sputnik* went into orbit on October 4, 1957.[24] When Vanguard crumpled back onto its launchpad in a fireball after rising only a few
feet off the sand at Cape Canaveral on December 6, a nationwide television
audience carried away an indelible image of embarrassing inferiority.

To the rescue came Wernher von Braun's Jupiter-C on January 31,
1958, when *Explorer I* joined *Sputnik* in orbit. The I-told-you-so sweet-
ness of his triumph appealed to the spirit of Everyman, vaulting him from
mere TV-star to national hero. Although the army still kept him on a
leash, warning him in June 1958 not to join the American Association
for the Advancement of Science because it was on the so-called pink list
of Communist-influenced organizations, his persona now had a life of its
own.[25] Some matters remained sensitive, however. In December, an assistant answered a query from a researcher at the University of Texas about
whether von Braun had ever been a member of the Nazi Party with a curt
denial: "In answer to your question, Dr. von Braun was not a Nazi."[26]

The satellite contest convinced Ike that the nation should have a civilian
space agency, which became von Braun's first nonmilitary employer.

Though the Soviets scored another goal when they launched the first man into space on April 12, 1961, it was von Braun's reliable Redstone that answered for the home team on May 5. The young new President Kennedy then decided to go for broke. From these Olympian heights, von Braun would not descend until after the moon was covered with boot prints. The past was erased. Nothing else mattered. He was the prophet of the Space Age.

NOTES

INTRODUCTION

1. I am indebted to John Forge's formulation in his essay "Responsibility and the Scientist" in *Science, Technology and Society* (Cambridge University Press, 1998), a text I have used with pleasure for years in my course about the social and political influences on science and technology at Johns Hopkins University.

1. A JUNKER'S LIFE

1. Franklin D. Roosevelt Inauguration address, March 4, 1933.
2. "Country folk do not simply supply food," wrote the dean of German agricultural economists, Max Sering. "They are the nation's energy reserve; all classes are provided by them with the physically and morally healthiest elements produced by breeding in tightly knit families in close touch with each other."
3. David Blackbourn and Richard J. Evans, eds., *The German Bourgeoisie* (Routledge, 1993), p. 225.
4. Ibid., pp. 246–47.
5. Genealogy is from Erik Bergaust, *Wernher von Braun* (National Space Institute, 1976), Appendix 1, which posits that the name derived from Braunau, the seat of Lueben County, about 200 miles southeast of Berlin. The family patriarch is believed to have lived there from 1285 to 1291. Like many family trees, this one contains tenuous branches that give it more the nature of a wishful pedigree— Wernher von Braun's father, in his autobiography, alluded to their relatively recent arrival in East Prussia, with estates acquired during the upheavals of the Napoleonic era. Membership in the lesser nobility often had a *burgerlich* (middle-class) element.

6. "Physically he happened to be a perfect example of the type labeled 'Aryan Nordic' by the Nazis," recalled Willy Ley in a reminiscential letter published in the *Journal of the British Interplanetary Society* 6 (June 1948). In Richard Ellmann's biography of Wilde, Lord Douglas—youngest son of the Marquess of Queensberry—is described as being "slight of build" and having "a pale alabaster face and blond hair." In temperament, Ellmann adds, Douglas was "totally spoiled, reckless, insolent, and, when thwarted, fiercely vindictive." Ellmann, *Oscar Wilde* (Vintage Books, 1988), p. 324. The fact that von Braun's female schoolmates would compare him to such an infamous (in those days) homosexual as Douglas, who had been a principal figure in Wilde's sensational libel trial in 1896, seems a very pointed barb.

7. Ley, letter, *Journal of the British Interplanetary Society.*

8. The phrase is from Goebbels's diary, as cited in Joachim C. Fest, *Hitler* (Mariner Books, 2002), p. 367. Estimates of the number of marchers ranged between 20,000 by hostile sources to 700,000 by pro-Nazis (Richard J. Evans, *The Coming of the Third Reich* [Penguin Press, 2004], p. 310).

9. Evans, p. 312.

10. Karl Dietrich Bracher, *The German Dictatorship* (Praeger, 1970), p. 214. Hitler's personal rise from hierophantic demagogue to statesman has been the subject of wave upon wave of scholarship and is still an active minefield for historians. See, for example, John Lukacs's *The Hitler of History*. (Some younger academics insist today on using the term National Socialist instead of Nazi, believing the latter to be such a powerful epithet by now that it obliterates degrees of participation in Hitler's movement. *Nazi* was of course an early abbreviation of *Nationalsozialistische*, sometimes also *Nazl*, both being diminutives of Ignatz and both a Bavarian slang name comparable to "Buddy.") Of pertinence here is not the precise course of the era's disasters, but the surrounding conditions that would have been impossible for any cogent German to ignore.

11. Roosevelt, Inauguration address. Language uncomfortably close to Nazi rhetoric, at least for the sensibility of today's reader, can be found elsewhere in mainstream American political statements from the interwar era. For example, Calvin Coolidge wrote a month before taking office as vice president: "Biological laws tell us that certain divergent people will not mix or blend. The Nordics propagate themselves successfully. With other races, the outcome shows deterioration on both sides. Quality of mind and body suggests that observance of ethnic law is as great a necessity to a nation as immigration law" (in "Whose country is this?," *Good Housekeeping* 72, no. 2 [February 1921]).

12. Street violence was an everyday occurrence. Public political murders, put at forty-two in 1929 and fifty in 1930, quadrupled in the first half of 1931 and kept increasing. At the same time, Nazis were claiming that they alone had the willpower to restore order (Peter Calvocoressi, Guy Wint, and John Pritchard, *Total War* [Pantheon, 1989], p. 53). Violent attacks against Jews reported by foreign newspaper correspondents were dismissed as "atrocity propaganda" by Nazi officials (Evans, p. 433).

13. Bracher, p. 253.

14. The Nazis did not invent concentration camps, which had been used by British forces to detain civilians under harsh conditions during the Boer War, but they had planned them since the early 1920s. A conservative estimate put the number of arrests in 1933 at more than 100,000 and deaths in custody at 600 (Evans, pp. 346, 348).

15. Fest, pp. 408–10. The Nazi Party was not a party in the ordinary sense of the word, since it was never satisfied with partial allegiance or partial dominance. It presented a comprehensive, total way of life. While membership was limited, the outer circle of sympathizers was gradually enlarged to create a mass movement that, like the party members but less intensively, was attached to Hitler personally and thus had a stake in his success. See also Andreas Dorpalen, *Hindenburg and the Weimar Republic* (Princeton University Press, 1964), pp. 463–69.

16. This so-called Enabling Act, which needed a two-thirds majority of the Reichstag, passed because the large Catholic Center Party voted for it at the Vatican's bidding. The Communist deputies were all in jail and could not vote.

17. "The flag high! The ranks tightly closed!" The first line of the song written by Horst Wessel, son of a prominent Lutheran clergyman in Berlin. A violent young convert to the Nazi movement, he was murdered in 1930 and lionized by Goebbels's propaganda office into a martyr. The verse continues: "SA marches with a firm, courageous pace. / Comrades, shot dead by Red Front and Reaction / March in spirit within our ranks."

18. The Weimar constitution was never abrogated—it was simply ignored—and the machinery of state was left largely intact, but power passed to numerous Nazi Party agencies with overlapping functions. The net result was that many decisions could be made only by the führer.

19. Fest, pp. 415–16. Of course, physical terror was one of Hitler's principal political weapons. Sadistic thugs were given license with the intention of securing his hold over Germany.

20. The state of Prussia, which was abolished in February 1947 by the Allied Control Commission, has an abstruse territorial history. At the turn of the eighteenth and nineteenth centuries, before the map of Europe was changed by the Napoleonic Wars, Prussia comprised almost more Polish than German land. After World War I, the Treaty of Versailles established the modern Polish state, forcing the cession of most of the Prussian provinces of Posen and West Prussia, as well as of one East Prussian county. After World War II, all former Prussian land east of the Oder River became part of Poland or the Soviet Union. On today's maps, few of the old German place names remain. For example, the medieval Prussian capital of Königsberg was virtually leveled during WW II and replaced by the austere Soviet architecture of Kaliningrad, home port of the USSR's Baltic fleet. It is now part of a sliver of Russian soil between Poland and Lithuania.

21. Blackbourn and Evans, p. 224.

22. Robert N. Proctor, *Value-Free Science?* (Harvard University Press, 1991), p. 236.

23. Details of Magnus von Braun's youth are taken largely from his autobiography, *Von Ostpreussen bis Texas* (Stollhamm, 1955). Other biographical details are from personal history statements compiled by American military authorities at the time of his immigration to the United States, obtained through Freedom of Information Act request.

24. "Lerne sparsam umbgehen mit diese Sachen, den es seind knape Zeiten," von Braun, p. 38. Land speculation and agricultural depressions that threatened the Prussian nobility fed a deep fear of Jews purchasing estates. The post-Napoleonic reform laws that had created provincial advisory councils did not allow Jewish owners of noble estates (who were, in any case, a tiny minority) to be members (Robert M. Berdahl, *The Politics of the Prussian Nobility* [Princeton University Press, 1988], p. 211).

25. Kant's philosophy appealed to those who wanted to isolate science from morality, to keep science and religion intact without concern about contradictions. Kantian dualism thus served to insulate science from political criticism (Proctor, pp. 79–80).

26. Berdahl, pp. 20–21. The Reich created in 1870 by Bismarck was a federation of German-speaking states dominated by Prussia. Widely regarded as the German nation, it was instead more like a consolidation of class power, with many Germans excluded from effective political activity.

27. Thus Hjalmar Schacht, Reichsbank president and Hitler finance minister who was acquitted of war crimes at Nuremberg, describing his life in *My First Seventy-six Years* (Allan Wingate, 1955).

28. *Hoffähigkeit*, or the status to appear at royal court functions, was granted to military officers beginning with the rank of lieutenant. Non-noble civil servants had to attain Class 4, equivalent to an assistant minister, to enjoy the same privilege.

29. See Gordon Craig, *The Politics of the Prussian Army, 1640–1945*, chapter 2. Hereditary serfdom was not abolished until 1807. Local government in cities was instituted a year later.

30. "Das liegt nicht an Kant—das liegt an mir; ich mus auf Erfahrungen aufbauen" (It was not Kant's fault—it was mine; I needed more experience) (von Braun, p. 33).

31. Ibid., p. 70. International Jewish solidarity, cast in terms of a conspiracy against gentiles, was of course part of the boilerplate of anti-Semitism.

32. Ibid., p. 71. Magnus's metaphor here of sheep and strong shepherd, combined with his phobia of democracy, makes him sound Nietzschean. But he was never a nihilist of any stripe.

33. Ibid., p. 72.

34. Goethe wrote: "Du musst steigen oder sinken, / Du musst herrschen und gewinnen / Oder dienen und verlieren, / Amboss oder Hammer sein" (Man must rise or fall, / He must win and rule / Or serve and lose, / Be the anvil or the hammer). The average lifespan in Germany at this time was less than fifty years.

35. Hajo Holborn, *A History of Modern Germany, 1840–1945* (Princeton University Press, 1982), pp. 405–6. I am especially indebted to this succinct and elegant reference.

36. Delbruck (1856–1921), who served as Vice Chancellor under Theobald von Beth-

mann-Hollweg during World War I, was known for moderation and social-mindedness as a minister of the monarch. Before the war he was a reserve captain and appeared often in uniform, despite his civilian job. After the war he played a prominent role as a representative of the German Nationalist People's Party, a right-wing group that voted against the Weimar Constitution and grew increasingly radical through the 1920s to become an important ally of the Nazis'.

37. When the Nazis came to power, Hermann Göring took over the mansion and turned it into what Magnus von Braun called a *Ritterburg*, or knight's castle, thus reversing the fashionable modernism it had become known for.

38. Von Braun, p. 79.

39. In reality, Kaiser Wilhelm's vulgar taste kept all the important modern art away from state galleries and theaters. The arts and literature were becoming more independent of the court than they had ever been before in Germany (Holborn, p. 404). Wilhelm, the first son of Queen Victoria's eldest daughter, had become emperor in 1888 at the age of twenty-nine. He was a spoiled flamboyant who stumbled through numerous personal and political scandals.

40. Von Braun, pp. 82–83. He added that "communism was a brother of democracy, with its enslaving of individuals."

41. Holborn, pp. 333–34.

42. Emmy was an orphan. Her mother had died in 1903 and her father in 1908. Personal History Statement dated November 26, 1947.

43. The Polish national movement in Germany had been centered in Posen for more than a century. Over the years, depending on the tone of Prussian Army administration, relations between Germans and Poles in the region varied from tense to murderous. German civil servants did not make the good impression on Poles that their Russian counterparts did, Magnus believed; first, because of "race and language" differences, and also because though the Russians were more "brutal," they knew how to drink with the Poles. "Good German qualities" such as order were not respected by the Poles, who saw the German virtue of discipline as lack of freedom.

44. Hans Roos, *A History of Modern Poland* (Knopf, 1966), pp. 3–5.

45. Magnus does not mention his second son's birth in the course of his autobiography. The book was intended to project his public self, not his personal life. As such, it contains virtually no reflection upon the feelings of people close to him and is nearly devoid of family matters. While this might be a function of his generation's formalism, it would seem in this case unusually defensive.

2. MEMORIES OF DEFEAT

1. In 1525 East Prussia became a duchy in fief to Poland, though Königsberg remained distinctly Germanic. In 1701 Königsberg became the seat of the new Prussian monarchy. Between 1758 and 1763 the Russian Army occupied the region. Napoleon's conquest came some forty years later.

2. Albert S. Lindemann, *Esau's Tears* (Cambridge University Press, 1997), p. 392.

3. As it turned out, many local Poles were enthusiastic about the declaration of war, despising the Russians even more than the Germans.

4. Leaders in Berlin were intent on doing nothing that might cause anger in London, where they had just completed a cordial set of colonial agreements. England would remain neutral, they presumed, provided Russia would appear as the aggressor. But if Germany invaded France through neutral Belgium as called for in the classic plan formulated by Prussian chief-of-staff Count Alfred von Schlieffen in 1905, British entry into the conflict was all but inevitable (Hajo Holborn, *A History of Modern Germany, 1840–1945* [Princeton University Press, 1982], p. 418, and Edgar Feuchtwanger, *Imperial Germany 1850–1918* [Routledge, 2001], p. 176). Even many years after the war, Magnus von Braun clung to the opinion that Germany had been squeezed into the conflict by Austria-Hungary, when in fact the German government had urged Vienna to go to war against Serbia and gave the strongest assurances that Germany would protect Austria-Hungary if intervention led to a broad conflict. All in all, the German government of 1914 reflected both the lack of political realism and the isolation of national thinking that characterized the country's ruling class.

5. S. L. A. Marshall, *World War I* (Houghton Mifflin, 1987), pp. 95–97; and Holborn, pp. 422–36. For years the Germans had fortified towns around the southeastern shoulder of East Prussia, but the area was weakly garrisoned and field commanders were authorized in the event of war to fall back west of the Vistula if necessary, thereby surrendering all of East Prussia. In the summer of 1914, the German espionage network of Polish Jewish agents quickly evaporated, making Russian intentions a matter of guesswork. This was just as well for the Russians, whose forces looked impressive on paper but suffered from too rapid mobilization (Robert B. Asprey, *The German High Command at War* [William Morrow, 1991], pp. 57–59).

6. See Robert Wohl, *The Generation of 1914* (Harvard, 1979), especially chapter 2, "Germany: The mission of the young generation," for the context of Ernst Jünger's calling August 1914 "the holy moment."

7. Russian invasion plans had been known to the Germans since 1902, when a Russian colonel had been bribed to reveal them. Various responses were practiced in annual maneuvers. There was clearly a certain amount of theatrics on both sides about all of this.

8. James L. Stokesbury, *A Short History of World War I* (William Morrow, 1981), pp. 61–67. Typical of the ingrown officer corps, Prittwitz was Hindenburg's wife's cousin. Hindenburg's chief of staff was Erich Ludendorff, who like Hindenburg would become a major figure in postwar politics. Though Prussian by birth, Ludendorff was not a Junker but the son of a poor tradesman. He more than made up for what was missing in his pedigree with talent, ambition, and industry. But he was inflexible, icy, prone to rages, and headstrong for war—all perhaps caricatures of the bona fide Junkers around him. Hindenburg's iron composure often served to restrain the *enfant terrible* Ludendorff. On the Hindenburg and Ludendorff

dynamic, see Andreas Dorpalen, *Hindenburg and the Weimar Republic* (Princeton, 1964), pp. 8–11.

9. Within days after the battle, his name became a household word. Hindenburg's previous military career had been honorable but not especially distinguished. What his superiors, and finally the entire nation, valued in him most were his appearance and temperament, which seemed to resonate with some cultural ideal. Of monumental stature, slow-moving, imperturbable, with a massive square head that conveyed unquestioned authority, he exuded a steadfast confidence that more than compensated for lack of initiative and imagination. The emergence of a Hindenburg myth, fostered by official propaganda, had mass psychological advantages as the German people began to feel nervous about the war's outcome (Dorpalen, pp. 8–11).

10. This was Hindenburg's first great defeat, a loss of some 40,000 men.

11. He does not indicate in his autobiography whose money this was—the local bank's, local depositors', or his own. In any case, it was a risky mission in wartime and a strange task for an attractive young woman, even if she was the Landrat's wife. An imposing figure, nearly six feet tall with clear blue eyes and blond hair, she would have been hard to intimidate. Typically, he offers no personal reflection on these matters.

12. Fritz Haber was the founding director, in 1911, of the Kaiser Wilhelm Institute for Physical Chemistry. His wife committed suicide just before the first successful use of battlefield gas at Ypres, France, in April 1915. In 1919 the Swedish Academy ignored the fact that the Allies considered Prof. Haber a war criminal and awarded him the Nobel Prize, focusing on his development of a seminal process for manufacturing ammonia from atmospheric hydrogen and nitrogen, despite the fact that his work allowed Germany to mass-produce explosives after the Allies blocked shipments of nitrate from Chile. Haber, a fervent patriot, left Germany when the Nazis came to power in 1933, because he was Jewish. He died the following year in Switzerland.

13. A survey by the Ministry of Interior in August 1914 had already revealed that the average farm had a stockpile of raw materials that would last for only six months. (Gordon Craig, *Germany 1866–1945* [Oxford University Press, 1980], p. 354). With the declaration of war, the military had taken over civilian government, including the organization of food supply.

14. City dwellers largely approved of food rationing, but farmers resented price ceilings on their goods, especially since industrialists were not being made to support the war through taxes on profits. Access to the limited supply of foodstuffs meant survival, already fueling a lively black market. Emmy may have been more sensitive than her husband to the danger of his probing too closely into local production.

15. Holborn, p. 449. Yet when the German and Austrian governments proclaimed a new kingdom of Poland in November 1916 under their military rule, the majority of volunteers for its army were Jews, perhaps an indication of how German civilization was perceived as less anti-Semitic than the Slavic cultures of Poland and Russia (ibid., p. 456).

16. For all practical purposes, Germany did not reach a state of total mobilization until 1917. Integration of wartime social and economic policies was actually never accomplished (Holborn, pp. 459–63).

17. Before the war, Germany had depended on imports for one-third of its food supply. By 1916 only 1350 calories daily in rationed food were available per person, with 2250 calories deemed necessary on average. During the "turnip winter" of 1916–17, the mass of nonagricultural population subsisted on that sour root. Anyone who could not afford unrationed food at obscene prices slipped into a starvation diet. About 750,000 people died of hunger. The grain harvest of 1917 was only half that of 1913 (Holborn, p. 460).

18. The potato crop in 1916 amounted to 25 million tons, compared to 54 million tons in 1913 and 1915, and remained far under normal for the rest of the war. A fodder shortage led to a sharp drop in meat production (Holborn, p. 460).

19. Richard Bessel, *Germany after the First World War* (Oxford University Press, 1993), p. 40. The urban working class was hardest hit by food shortages.

20. Feuchtwanger, p. 182. In October 1916, the Prussian minister of war ordered a "Jew count" in the army, pandering to the belief that Jews were profiteers and avoided military service.

21. Berne, Switzerland, of course, a notorious spies' nest in a neutral country.

22. Verdun was the ultimate horror, in which about 700,000 died on both sides (Feuchtwanger, p. 183).

23. Helfferich, a member of the Conservative Party, had risen to prominence as an architect of the Deutsche Bank's financial support of the Bagdad Railway project in Turkey. He was perhaps best known for switching from fierce opposition to enthusiastic support for unrestricted submarine warfare. He consistently opposed taxation to pay for the war. "We hold fast to the hope of presenting our opponents at the conclusion of peace with the bill for this war that has been forced upon us," he said in 1915 while serving as secretary of the treasury. After the war he joined the Deutschnationale Volkspartei (DNVP)—the party of Wolfgang Kapp—and engaged in the right-wing nationalist demagogy later adopted by the Nazis (Craig, *Germany*, pp. 365–57; and Feuchtwanger, p. 72).

24. All existing stocks of ammunition had been used up by October 1914. After the Battle of the Somme (July–November 1916), the German economy began to be stretched to, and ultimately beyond, its limits.

25. This view was widely shared, as unwise compromise dogged Bethmann's career. Von Braun and other Junker bureaucrats probably felt invidiously superior to him, as well. He had been born in 1856 to an *haute bourgeois* family that owned a vast East Prussian estate. Intelligent, cultured, well-educated, he had risen to the top despite being not quite a true aristocrat.

26. Holborn, p. 450.

27. Ralph Haswell Lutz, ed., *Fall of the German Empire, 1914–18*, vol. 1 (Stanford University Press, 1932), p. 412.

28. As a food administrator, Michaelis had instituted the grain and bread rationing

office in the Interior department, one of the more efficient of these stumbling agencies. It seems likely that von Braun had made his acquaintance there.

29. John G. Williamson, *Karl Helfferich* (Princeton, 1971), p. 231.

30. When Hitler came to power, he turned the *Reichpressechef* agency into the Ministry of Propaganda under Goebbels. "It is the same for parents," von Braun quipped. "You never know how your children will turn out."

31. Magnus von Braun, *Von Ostpreussen bis Texas* (Stollhamm, 1955). p. 142.

32. "Politically, he's just a child," von Braun wrote with amazing audacity.

33. Von Braun, p. 145.

34. During 1917–18, the military resorted to searching farms for hidden food reserves (Bessel, p. 37).

35. Holborn, p. 502.

36. Wilhelm II went by special train into exile in Holland on November 10, 1918. Ludendorff's flight to Sweden was less dignified, disguised as he was in blue spectacles and false whiskers. Hindenburg remained to take supreme command of the army.

37. *The New York Times*, November 11, 1918.

38. Bessel, p. 69.

39. Robert L. Hetzel, "German monetary history in the first half of the twentieth century," *Economic Quarterly*, January 1, 2002.

40. Sparticist leader Rosa Luxemburg was lynched in Berlin by *Freikorps* officers on January 15, 1919. General Ludendorff returned to Germany unmolested in February.

41. Ian Kershaw, *Hitler 1889–1936: Hubris* (W. W. Norton, 2000), p. 110–16.

42. The Weimar constitution was written by a Jewish legal scholar, Hugo Preuss, and many of the leading officials were Jewish (Lindemann, pp. 481–83).

43. The revolution that was occurring was primarily a military mutiny. The radical soldiers' councils that had come into being in every unit were an expression of the army's disillusionment with its leaders, not of affinity for Bolshevism. The retreat of German forces was actually quite orderly, aided by dry weather. See John Wheeler-Bennett, *The Nemesis of Power* (MacMillan, 1954), pp. 22 and 27.

44. Kapp had been born to a family of German immigrants in New York in 1868. A fanatical reactionary, he returned to Germany and became head of a district agricultural office in East Prussia. He played a peripheral role in the intrigues that led to Bethmann-Hollweg's ouster as chancellor. His Fatherland Party had fulminated against electoral reform and compromise peace during the war. With 2500 local chapters and 1.25 million members, it amounted to the largest concentration of German chauvinism before the Nazis. After the war he was a member of the presidium of the *Deutschnationale Volkspartei*—a monarchist, Protestant party with strongholds in agricultural areas east of the Elbe.

45. A network of army corps commands, each controlled by an officer who was above the law, enforced a state-of-siege invoked by the kaiser at the beginning of the war to guarantee internal security and censorship. Von Braun knew that he could have

been charged with high treason, and found the monetary fine rather ridiculous under the circumstances, but hanging the Prussian nobility was still unthinkable. There was strong sympathy for Kapp across Prussia's eastern provinces, where three provincial presidents, three district commissioners, and 88 Landrate were removed for their support. In a purge of the Reichswehr after the putsch, only 172 officers (including 12 generals) were dismissed. Though "nonpolitical" in theory, the army was predominantly rightist and critical of the Weimar Republic (Holborn, pp. 585–87).

46. The debt burden of German agriculture had grown uncomfortably high in prewar decades as cheap American grain flooded world markets. The indebtedness of estate owners was especially high due to investment in modern tools and methods. Availability of credit and protection from usurious loans was therefore a central concern for Junker landowners.

3. "HIGHLY TECHNOLOGICAL ROMANTICISM"

1. The description of futilitarian postwar youth—most famously Ernest Hemingway and other expatriates in Paris during the 1920s—as a "lost generation" originated with Gertrude Stein, who in *The Autobiography of Alice B. Toklas* attributed the phrase *generation perdue* to the proprietor of an auto repair shop speaking to an inept mechanic who had served in the war. In *Good-Bye to All That*, Robert Graves called the war "a sacrifice of the idealistic younger generation to the stupidity and self-protective alarm of the elder." On the other hand, it is misleading to brand the whole postwar generation as unfit for normal civilian life. Some 11 million Germans served in the armed forces during the war and survived, but membership in the sociopathic *Freikorps* was at most about 400,000. The antiwar Reichsbund attracted double that number. See Richard Bessel, *Germany After the First World War* (Oxford University Press, 1995), p. 258.

2. Bessel, p. 273.

3. Bessel, p. v. American soldiers who served in the war received an illustrated certificate, signed by President Wilson, titled "Columbia Gives to Her Son the Accolade of the New Chivalry of Humanity," which portrayed a toga-clad woman resting a sword upon the shoulder of a kneeling infantryman. What the ravaged young men returning from St. Mihiel or the Meuse-Argonne made of this can only be imagined. Personal collection of the author.

4. Walter Dornberger, *V-2* (Viking Press, 1954), p. 29. Dornberger, an engineer, artillery officer, and veteran of the Great War, became the leader of German military rocket development during the Third Reich and arguably was of more importance than Wernher von Braun. Dornberger's public reputation as the general who fired V-2s at London and Antwerp made it impossible after the war to cast him as a visionary spaceflight hero, whereas von Braun's civilian status rendered him more presentable as such. Dornberger's book is widely acknowledged by historians to be invaluable, but those who have placed von Braun at center stage necessarily write

pejoratively about Dornberger. Michael Neufeld, for example, judged his memoirs to be "egocentric, unreliable, and, when it comes to the SS and concentration camp issues, dishonest." A similar case could be made for von Braun and other V-2 alumni, of course. See Michael J. Neufeld, *Von Braun: Dreamer of Space, Engineer of War* (Knopf, 2007), p. 496, note 13.

5. Dornberger wrote that the logical consequence of the Treaty of Versailles was that the German Army sought new weapons that would increase the power of the relatively small number of permitted troops without violating the treaty (*V-2*, p. 19). But treaty violations were not as much of a concern as he claimed, because they occurred repeatedly in other Reichswehr bailiwicks. He further mentioned that a report was made to the Minister of Defense, which near the end of 1929 decided to start researching new possibilities for military rocketry.

6. Martin Broszat, *Hitler and the Collapse of Weimar Germany* (Berg, 1987), p. 46.

7. "The Treaty of Versailles shocked me; it seemed destined to cause another war some day, yet nobody cared," Robert Graves wrote in *Good-Bye to All That*.

8. *The Treaty of Versailles and After* (Washington, DC: U.S. Government Printing Office, 1947), pp. 325–37.

9. Richard Bessel and E. J. Feuchtwanger, eds., *Social Change and Political Development in Weimar Germany* (Croom Helm Ltd., 1981), p. 99.

10. Hajo Holborn, *A History of Modern Germany, 1840–1945* (Princeton University Press, 1982), pp. 587–88. *Stahlhelm*, which also called itself the "League of Front Soldiers," founded at the end of 1918 as a patriotic veterans group, advocated liberation from "the chains of Versailles" and generally followed the political program of the German Nationalist Party. Other cadres, including the storm troopers (SA) of Hitler's National Socialist Party, were more radical descendants of Freikorps paramilitary units. The Reichswehr depended on the assistance of these irregulars—in forming frontier guards along the German-Polish border, for example—to skirt the treaty's limitations.

11. Launched with a speech in January 1918, Wilson's program for postwar order offered the central principle of national self-determination, which was attractive to a supine Germany. But when it was only applied to Germany's disadvantage, as in the deeply resented loss of Prussian territory to Poland, the American president became equated with trickery (Edgar Feuchtwanger, *From Weimar to Hitler* [St. Martin's Press, 1993], pp. 47 and 51).

12. Magnus von Braun, *Von Ostpreussen bis Texas* (Stollhamm, 1955), p. 187. It is unclear how much he earned from this position. Under the pressure of runaway inflation, Raiffeisen eventually collapsed, though it was saved by its primary investors. Later in the decade, he was an unpaid vice president of the bank. Presumably, throughout this time, he depended on income from family estates in Prussia.

13. It also kept him active in agricultural circles that would remain important to the economic and political development of the Reich throughout the 1920s, a decade when about one-third of the German population were still dependent on agriculture for their livelihoods. Bessel and Feuchtwanger, pp. 12–13. All in all, the

upheavals of the immediate postwar years did not succeed in sweeping the Junkers from positions of social and political power.

14. Cited in Jeffrey Herf, *Reactionary Modernism* (Cambridge University Press, 1990), p. 2. The phrase is from "Germany and the Germans," a speech given by Thomas Mann at the U.S. Library of Congress on May 29, 1945. Mann traced this intellectual spirit back to Bismarck's empire, with its "mixture of robust timeliness, efficient modernness on the one hand and dreams of the past on the other." In the same speech, he warned: "Wherever arrogance of the intellect mates with the spiritual obsolete and archaic, there is the Devil's domain." *Technik und Kultur* was the title of a journal for graduates of the technical universities.

15. Herf, pp. 221–22.

16. "We were very much in the position of aviation pioneers when the airplane could only be developed because of its military value" (Wernher von Braun, "Reminiscences of German rocketry," *Journal of the British Interplanetary Society* 15, no. 3 [May–June 1956], p. 130). This was a mistaken and self-serving reading of the pertinent history. Though the aviation pioneers realized from the start that flying machines had military value, and sought government funding on that basis, they eagerly pursued civilian applications. That the early development of airplanes coincided with World War I, as the discovery of atomic fission occurred just before World War II, reflects how war shaped important twentieth-century technologies.

17. In 2004, the first successful non-government-funded journey into space depended on the largesse of a single billionaire businessman.

18. Whiggist in the sense of Herbert Butterfield's landmark work in the historiography of science, *The Whig Interpretation of History* (1931): "to emphasize certain principles of progress in the past and to produce a story which is the ratification if not the glorification of the present." Perhaps unsurprisingly, the tendency has been strongest among historians employed by various United States government agencies with direct interest in a triumphal story, especially the National Aeronautics and Space Administration and the Air and Space Museum in Washington, DC. Much of the writing from the heyday of the Space Age has a propagandistic tone that today seems quaint and naïve.

19. See, for example, Neufeld, *Von Braun: Dreamer of Space, Engineer of War.* Based on an exhaustive reading of archival evidence by a longtime curator at the Air and Space Museum, this important work corrected much of the factual error enshrouding von Braun, while maintaining some of the heroic accoladia expressed by its title. In an attempt to smooth the way for von Braun's immigration to the United States after World War II, the phrase "mere opportunist" was first applied to him in a U.S. military security report dated February 26, 1948, that changed a September 18, 1947, report that had called him a "potential security threat." Documents via FOIA request. See also Linda Hunt, "U.S. Coverup of Nazi Scientists," *Bulletin of the Atomic Scientists*, April 1985.

20. See article defending von Braun by Ernst Stuhlinger in *The Huntsville* (Alabama) *Times*, October 14, 1995.

21. Hjalmar Schacht, *My First Seventy-six Years* (Allan Wingate, 1955), p. 176.

22. Biographical details from an interview with Oberth conducted by Martin Harwit and Frank Winter for the National Air and Space Museum, November 14 and 15, 1987. I am grateful to Michael Neufeld for providing the text of this and other such oral history interviews as part of my 1994–95 Verville research fellowship at the museum.

23. Oberth later said that his dissertation was rejected because "it dealt mainly with physical-medical subjects" (Frank H. Winter, *Rockets into Space* [Harvard University Press, 1990], p. 19). A year earlier, the Heidelberg faculty had awarded a PhD in Romantic literature to twenty-four-year-old Joseph Goebbels.

24. In July 1922, the British Consul in Frankfurt wrote: "[T]he educated classes, deprived, in a great majority of cases, of the right to live and bring up their families in decency, are becoming more and more hostile to the Republic and open in their adhesion to the forces of reaction. Coupled with this movement, a strong and virulent growth of anti-Semitism is manifest" (Gerald D. Feldman, *The Great Disorder: Politics, Economics, and Society in the German Inflation, 1914–1924* [Oxford University Press, 1997], p. 449).

25. "Such a piece cannot be published by our serious publishing house," Oberth recalled being told by the Viehweg company, to whom he had first offered his manuscript. He was advised not to publish it at all by Max Wolf, professor of astronomy at Heidelberg (see Harwit and Winter interview). Evidently never constrained by reality, he had sent an unsolicited proposal to the German War Department in 1917, proposing a long-range liquid-fuel rocket that was 82 feet tall and weighed 10 tons. Nearing a state of total collapse of its conventional forces, the War Department declined interest. See Winter, *Rockets into Space*, pp. 20–25, for a respectful discussion of Oberth's book.

26. Dornberger, p. 68.

27. Valier was described by a contemporary as "a sort of stunt merchant who fooled around with cars driven by solid rockets" (Letter to Walter Riedel from A. V. Cleaver, British Interplanetary Society council member, December 29, 1951; Imperial War Museum, German Misc. 148).

28. Real income of the individual in 1928–29 remained 6 percent less than prewar income. Unemployment during the winter seasons from 1923 to 1929 ranged from 1.5 million to 2.6 million in a workforce of about 30 million (Holborn, pp. 638–39).

29. Evans, p. 223.

30. The track had been built beginning in 1907 by an automobile club. It consisted of two parallel straightaways about 6 miles long (clearly suitable for a rocket car), joined at the ends by short curves. Today it is incorporated into the autobahn system around central Berlin, with old wooden grandstands preserved as historical monuments.

31. Valier was killed on May 17, 1930, by the explosion of a liquid-fuel rocket engine being developed for stunt cars at a Berlin factory, owned by Paul Heylandt, that

produced compressed and liquid gases for industry. Two men who would later be prominent in German military rocketry, Arthur Rudolph and Walter Reidel, took part in this work (Willy Ley, *Events in Space* [David McKay Company, Inc., 1969], pp. 28–29). See also Michael Neufeld, "Weimar culture and futuristic technology: The rocketry and spaceflight fad in Germany, 1923–1933," *Technology and Culture* 31 (1990).

32. Bruce Murray, *Film and the German Left in the Weimar Republic: From Caligari to Kuhle* (University of Texas, 1990), pp. 20–25.

33. Eric D. Weitz, *Weimar Germany: Promises and Tragedy* (Princeton, 2007), p. 106.

34. In June 1929, Hugenberg persuaded Hitler to join a committee against government acceptance of the Young Plan, which lowered World War I reparations payments (but did not eliminate them). The campaign failed, but Hitler gained wide exposure in the Hugenberg-controlled press (Ian Kershaw, *Hitler, 1936–45: Nemesis* [W. W. Norton, 2001], p. 320).

35. Alexandra Richie, *Faust's Metropolis* (Carroll & Graf, 1998), pp. 351–56. Larry Eugene Jones, "Alfred Hugenberg and the Formation of the Hitler Cabinet," *Journal of Contemporary History* 27 (1992), pp. 63–87. The party program of 1931, drafted under Hugenberg's influence, stated that "we resist the subversive, un-German spirit in all forms, whether it stems from Jewish or other circles. We are emphatically opposed to the prevalence of Jewdom in the government and in public life, a prevalence that has emerged ever more continuously since the revolution [1918]" (Richard J. Evans, *The Coming of the Third Reich* [Penguin Press, 2003], p. 95). In March 1933, UFA began a policy of cutting Jewish staff and actors (Evans, p. 405). Lang immigrated to France in April of that year, eventually resuming his career in Hollywood. His wife, the screenwriter Thea von Harbou, who had written *Frau im Mond*, stayed in Germany and supported Hitler.

36. Harwit and Winter interview.

37. According to Willy Ley, a founding member of the society, Valier had been warned by a lawyer after giving a lecture about space travel in Munich that soliciting money for Oberth's work should occur under the aegis of a legally chartered society, presumably to protect against charges of fraud. The *VfR* was thus founded in Breslau (now Wroclaw, Poland), where the court refused to grant a charter until the group defined what was meant by *Raumschiffahrt*, which was not in the dictionary (Ley, p. 24).

38. Letter from Nebel to Major Bedenschatz, dated August 23, 1933, in Imperial War Museum, MI 14/801 (V).

39. By January 1930, labor records indicated that about 14 percent of the working-age population were unemployed, but the true figure was probably closer to 20 percent (Kershaw, p. 318).

4. AN HEIR OF CREDIBILITY

1. German Army officer Walter Dornberger recalled many years later that when Magnus von Braun visited him in 1933 "he told me with frequent headshakings

that he had no idea where his son had acquired this strange technological bent" (Walter Dornberger, *V-2* [Viking Press, 1954], p. 27).

2. According to Walter Riedel's "Raketenentwicklung," an homage to Max Valier as a pioneer of rocketry, there were two German rocket groups in 1930. One consisted of Hermann Oberth, Willy Ley, Rudolph Nebel, Klaus Riedel (not related to Walter), and Wernher von Braun. Nebel became the leader of this group and was mentioned prominently in a January 2, 1933, Frankfurter Zeitung account of an amateur rocket launch at Magdeburg as the "Grunder und Leiter des Raketenflugplatzes." The other group was organized by Max Valier with money from Paul Heylandt, a manufacturer of industrial liquid-oxygen gear (Imperial War Museum, German Misc. 148).

3. *Berliner Morgenpost,* June 11, 1973; Deutches Museum.

4. Wernher von Braun, "Reminiscences of German rocketry," *Journal of the British Interplanetary Society* 15, no. 3 (May–June 1956).

5. Wernher von Braun, *Recollections of Early Childhood/Early Experiences in Rocketry, As Told by Wernher von Braun, 1963* (Marshall Space Flight Center).

6. Willy Ley, "Count von Braun," *Journal of the British Interplanetary Society* 6 (June 1948). How this visit was arranged was not explained. It is fair to say that Ley, a successful popular science writer in Germany before World War II and in the United States afterward, was responsible for spreading the Whiggist narrative of Weimar rocketry.

7. Interview with Oberth, NASM, 1987.

8. Erik Bergaust, *Wernher von Braun* (Washington, DC: National Space Institute, 1976), p. 40. Bergaust was editor of *Missiles and Rockets*, a trade magazine whose advisory board was heavily larded with German engineers.

9. Von Braun, "Reminiscences," and *Berliner Morgenpost*.

10. Department of Defense security questionnaire, undated. The French Gymnasium would certainly have been regarded as a cosmopolitan institution, socially and intellectually, though its classical curriculum was conservative. It accepted students whose parents could afford it, including Jews. The Hermann Lietz schools' reputation was based on pedagogical, not political, progressivism, and emphasized learning in a nonurban setting. Lietz (1868–1919) was a German educational reformer who founded several *Landerziehungsheime* (country boarding schools) based on the English Abbotsholme model of well-rounded, rather than strictly academic, students. Classroom study of traditional subjects was combined with outdoor recreation and instruction in manual arts such as carpentry. In 1928, Wernher moved to a new Lietz school on Spiekeroog, a then-remote island just off the North Sea coast, west of Bremerhaven. (If Oberth's comment about Wernher being "not very skillful with his hands" is accurate, then the expensive Lietz experience was somewhat wasted on him.) Goethe is said to have frequented the Ettersburg Castle during the 1770s while writing *Faust* and other works.

11. Bergaust, p. 35. In general, private schools in Germany were considered of inferior quality to the gymnasiums—places where students who had stumbled academically could try to regain their footing.

12. Von Braun, "Reminiscences." Another version of this anecdote is notable in retrospect for his comment that "it never occurred to me" that bystanders "were not prepared to share the sidewalk with my noble experiment," though he "yelled a warning and men and women fled in all directions" ("Space man—the story of my life," *American Weekly*, 1958).

13. Von Braun, *Recollections*.

14. The debate was over whether to raise employer contributions to unemployment insurance from 3.5 to 4 percent of wages. The Social Democrat Hermann Muller was replaced by Heinrich Bruning from the far right of the Catholic Party, a move facilitated by Major-General Kurt von Schleicher in the Defense Ministry (Ian Kershaw, *Hitler, 1889–1936: Hubris* (W. W. Norton, 2000), p. 323.

15. Hindenburg grudgingly abided by the Weimar constitution, but also decided that German diplomatic outposts and ships should fly the imperial black-white-red flag as well as the Republic's black-red-gold banner.

16. Richard J. Evans, *The Coming of the Third Reich* (Penguin Press, 2004), p. 248.

17. This would have amounted to about 10 percent of Germany's national income every year, a politically untenable figure even if fiscally possible (Gerard D. Feldman, *The Great Disorder: Politics, Economics, and Society in the German Inflation, 1914–1924* [Oxford University Press, 1993], p. 406).

18. Hajo Holborn, *A History of Modern Germany, 1840–1945* (Princeton University Press, 1982), p. 643.

19. Similarly expedient associations between respectable investors and eccentric inventors had been common during the early cottage-industry era of aviation.

20. Since 1946 called the Technical University of Berlin. Agent report submitted March 17, 1959.

21. "Even before the Nazis came to power in 1933, the Technical College became a hive of Nazi activity, particularly among its students. Many of the professors were also more than merely tolerant of the National Socialists" (From the section headed "Pillar of the Nazi War Machine" on the history page of the Technical University of Berlin Web site).

22. *Berliner Morgenpost*.

23. In the middle of October a strike was called by some 140,000 metal workers in Berlin to resist wage cuts. By November they accepted a 6 percent decrease (Martin Broszat, *Hitler and the Collapse of Weimar Germany* [Berg, 1987], p. 13).

24. Henry Ashby Turner, *German Big Business and the Rise of Hitler* (Oxford University Press, 1985).

25. Von Braun, "Reminiscences."

26. Von Braun, *Recollections*.

27. Holborn, p. 670. With 18.3 percent of the popular vote, riding a landslide especially strong in the Protestant countryside of northern and eastern Germany, the NSDAP was now the second largest party in the Reichstag behind the Social Democrats. The Communists also increased their share to 13.1 percent (Kershaw, pp. 333–34).

28. *Berliner Morgenpost*.

29. Von Braun, *Recollections*.
30. Von Braun, "Reminiscences."

5. CHILDHOOD'S END

1. Detlev J. K. Peukert, *The Weimar Republic* (Hill and Wang, 1992), p. 252. In 1932, the suicide rate in Great Britain was 85 per million inhabitants, 133 in the United States, and 260 in Germany (ibid., p. 280). See Eric W. Hobsbawm's reminiscence of the last year of Weimar in *The London Review of Books*, January 24, 2008, pp. 34–35.

2. Hajo Holborn, *A History of Modern Germany 1840–1945* (Princeton University Press, 1982), pp. 673–83. "It was the outbreak of the banking crisis in the summer of 1931 that made the German depression so severe. . . . [T]he collapse of the banks in central Europe had a major social, psychological and political impact. Capitalism appeared to have crashed with the banks, and this helped to discredit existing political systems" (Harold James, *The German Slump: Politics and Economics 1924–1936* [Clarendon Press, 1986], pp. 283–84).

3. *Vossische Zeitung*, October 2, 1930. Quoted in Martin Broszat, *Hitler and the Collapse of Weimar Germany* (Berg, 1987), p. 16.

4. Politicians committed to the Republic seemed to die young: Friedrich Ebert at 54, Gustav Stresemann at 51, Walter Rathenau at 55 (assassinated), Matthias Erzberger at 46 (assassinated).

5. Broszat, p. 86. "I could well understand the appeal of both [nationalism and Nazism] to German boys," recalled Eric Hobsbawm, *London Review of Books*.

6. Broszat, p. 67. Goebbels was put in charge of overall propaganda in 1930 (ibid., p. 79).

7. Hitler appointed von Schirach (1907–1974), son of a Weimar theater director and an American socialite, to lead the Hitler Youth Organization in October 1931. He eventually oversaw the deportation of 185,000 Jews from Vienna to concentration camps in Poland and was sentenced by the Allies to twenty years in Spandau prison as a war criminal (Nuremburg trial proceedings, vol. 14, May 23, 1946).

8. Richard J. Evans, *The Coming of the Third Reich* (Penguin Press, 2004), p. 260.

9. Holborn, p. 603.

10. The von Braun family's Berlin home addresses in 1932 were listed as Friedrich Ebert Strasse 19 and Wilhelmstrasse 71 on a "Personal History Statement" signed by his father on November 26, 1947, in conjunction with immigration to the United States (FOIA request). At the glider school, von Braun befriended another student pilot his age, Hanna Reitsch (1912–1979), who would go on to become a celebrity aviator in Germany as popular as Amelia Earhart in the United States, as well as an important military test pilot and personal friend of Hitler's until his last days. The arc of her career was strikingly similar to von Braun's; they maintained their friendship for the rest of their lives. Gliding at this time was a sport that symbolized resistance to Treaty of Versailles restrictions on German aviation. See Peter Fritz-

sche, *A Nation of Flyers: German Aviation and the Popular Imagination* (Harvard University Press, 1994).

By 1925, there were fewer than 10,000 estate owners in all of Germany, but though comprising less than one-half of 1 percent of landowners, they owned over 22 percent of the cultivated land. Both Junkers and peasants were now integrated into a capitalist system, but the privileged position of the estate owners (many of whom were incompetent at farming)—who had substantial control over the personal and political lives of their employees—still made them leaders and spokesmen for German agriculture (David Abraham, *The Collapse of the Weimar Republic: Political Economy and Crisis* [Holmes Meier, 1988], pp. 45–47). The Great Depression came earliest to the agrarian sector. In 1930, some 62,000 acres of farmland were sold at compulsory auctions in East Prussia alone. From the summer of that year onward, the Nazis were the leading agrarian movement (Broszat, pp. 13 and 73). Silesia was not annexed by Prussia until 1742. Although it was inhabited by many "grand seigneurs" and industrial magnates, it had a large number of small, poor, socially self-conscious and politically active Junkers (Lysbeth Walker Muncy, *The Junker in the Prussian Administration under William II, 1888–1914* [Brown University Press, 1944], p. 34). The Oberwiesanthal estate is today in Polish territory.

11. "He told me that he wanted to get his doctor's degree in engineering from the Swiss Federal Technical College in Zurich," remembered Willy Ley in 1948, without saying why ("Count von Braun," *Journal of the British Interplanetary Society* 6 [June 1948]).

12. Matthias Schmidt, *Albert Speer, the End of a Myth* (St. Martin's Press, 1984), p. 29.

13. Constantine D. J. Generales, "Wernher von Braun," *New York State Journal of Medicine*, November 1977, p. 2174.

14. Von Braun, "Reminiscences," p. 128.

15. Constantine Generales, "Recollections of Early Biomedical Moon-mice Investigations," *Smithsonian Annals of Flight*, no. 10 (1974), p. 77.

16. Generales, "Wernher von Braun," p. 2174. Wernher von Braun, "Reminiscences of German rocketry," *Journal of the British Interplanetary Society* 15, no. 3 (May–June 1956), p. 128.

17. Generales eventually acquired his MD from the University of Berlin and went on to be named "coordinator of space medicine" at New York Medical College. He died in 1988 at the age of 79 (*The New York Times*, June 11, 1988).

18. Von Braun, "Reminiscences," p. 128.

19. "The motor was located in the nose, not for any scientific reason, but simply because Nebel had scrounged a truckload of aluminum tubing which could only be used if the motor dragged the tanks by the fuel lines," von Braun wrote many years later (von Braun, *Recollections*).

20. The Vernesque 1897 novel about a trip to Mars, *Auf zwei Planeten* by Kurd Lasswitz, was mentioned by Willy Ley in *Events in Space*, p. 5.

21. Copy from Imperial War Museum, file MI 14/801 (V). This *Waffenamt* file labeled "Nebel" is actually Major Walter Dornberger's personal file.

22. In 1915 and 1916, Goddard approached the U.S. Naval Consulting Board and the Smithsonian Institution for funding, emphasizing military applications to the first and scientific to the second. In 1925 and 1926, his small liquid-fueled rockets weighing a few kilograms flew for distances of about 10 meters. In July 1929, he attracted publicity and the philanthropic introductions of Charles Lindbergh with a 3.5-meter-long rocket that rose to an altitude of 30 meters. By 1932 he was spending $2000 per month to develop high-altitude rockets that never came to fruition (David H. DeVorkin, *Science with a Vengeance* [Springer-Verlag, 1992], pp. 7–11).

23. See Becker letters from IWM file cited in note 21. A summary of the army's six-year relationship with Nebel is dated January 11, 1937. In 1931, Becker's ordnance section also funded research at the Heylandt Works (*Aktiengesellschaft für Industrie Gas Verwertung*), which produced industrial gases, that was a continuation of Max Valier's rocket stunt car development. Valier had been killed by an exploding rocket motor there in 1930.

24. Holborn, pp. 688–89.

25. Police in Prussia counted 155 killed in political clashes in 1932, with 82 deaths and 400 injuries during 461 riots in the first seven weeks of the campaign (Evans, p. 270).

26. Papen was a former cavalry officer who became military attaché at the German Embassy in Washington, DC, during World War I, a post from which he was expelled in 1916 by the American government because of sabotage activities. A member of the Catholic Zentrum party due to his religion, he served for ten years in the Prussian parliament, gaining notoriety by calling for a dictatorship in 1923. He split from the party in 1925 by supporting Hindenburg against the Zentrum candidate. Though of elegant manner and a majority shareholder in the party's Berlin newspaper, he was never taken seriously politically (Holborn, p. 693; Evans, p. 283).

27. Holborn, p. 693.

28. Magnus von Braun, *Von Ostpreussen bis Texas* (Stollham, 1955), p. 208.

29. As a member of the DNVP, he adhered to monarchist principles. "Democracy is a form of government in which it is permitted to wonder aloud what the country could do under a first class management," he quoted from the American "Senator Soaper" syndicated humor column (ibid., p. 209). Since 1930 he had been vice president of the *Reichsverbands der Landwirtschaftlichen Genossenschaften*, an agricultural cooperative association.

30. Abraham, p. 98.

31. See Horst Gies, "The NSDAP and agrarian organizations of the Weimar Republic," in *Nazism and the Third Reich*, Henry Turner, ed. (Quadrangle Books, 1972).

32. Von Braun, "Reminiscences," p. 129.

33. Dornberger summed up the army's intentions in his 1954 memoir: "We wanted to have done once and for all with theory, unproved claims, and boastful fantasy, and to arrive at conclusions based on a sound scientific foundation. We were tired of imaginative projects concerning space travel." That is, of course, they wanted the

practical information necessary to mass-produce weapons (Dornberger, pp. 20–21).

34. There is evidence from the Russian State Military Archive that one of the Raketenflugplatz group, Rolf Engel, had Communist sympathies. In 1932, he volunteered to Soviet military authorities a report on German rocketry and even offered to bring a group of specialists to the Soviet Union. See Mark Harrison, "A Soviet quasi-market for inventions: Jet propulsion, 1932–1946," *Research in Economic History*, vol. 23 (Elsevier, 2005), p. 29. This article also cites archival evidence of an internal Red Army report from May 1931 on rocketry abroad, mainly in Germany and the United States: "A number of German research groups and firms, including Junkers and Opel, were described as competing for patents and funding under the umbrella of a voluntary society for space travel including armed services representatives."

35. Von Braun, "Reminiscences," p. 129.

36. "It was our job to separate the wheat from the chaff, and that was no small task in a sphere of activity so beset with humbugs, charlatans, and scientific cranks, and so sparsely populated with men of real ability" (Dornberger, p. 29).

37. Almost 25 years later, von Braun stated that "it seemed that the funds and facilities of the Army were the only practical approach to space travel" ("Reminiscences," p. 130).

38. Von Braun, "Reminiscences," p. 129.

39. Ibid., p. 130.

40. "Exactly how he proposed to bridge the gap between our small, sputtering rocket and a huge, passenger-carrying spaceship wasn't quite clear to us, nor, it may be supposed, to him. Full well he must have known that it would require a truly vast expenditure" (von Braun, "Reminiscences," p. 131).

41. "I had been struck during my casual visits to Reinickendorf by the energy and shrewdness with which this tall, fair, young student with the broad massive chin went to work, and by his astonishing theoretical knowledge." As soon as von Braun began at Kummersdorf, Dornberger hired a skilled mechanic who had worked at the Raketenflugplatz (Dornberger, p. 27).

42. Michael Neufeld, *Von Braun* (Knopf, 2007), p. 53.

43. Von Braun's 1963 *Recollections* says incorrectly that he graduated in the spring of 1932 with a bachelor's degree in aeronautical engineering. Friedrich-Wilhelm, founded in 1810 as the University of Berlin, has been called Humboldt University since 1949. In Walter Dornberger's memoir, he refers to von Braun at this stage as "our nineteen-year-old 'student,'" apparently using quotation marks around *student* to convey that this was not quite the right word for his position as a member of Dornberger's specialist staff (Dornberger, p. 26).

44. Ley, "Count von Braun."

45. von Braun, "Reminiscences." In this and other versions of essentially the same memoir, including *Recollections*, which were drafted in hindsight and probably composed for him in colloquial English by U.S. Army public relations staff, von Braun insisted that space travel was always his main purpose, that the German

army was only milked as a golden cow in order to pursue this objective, that the morality of working for the military and the possible abuse of his inventions were never considered. "Our minds were always far out in space," Willy Ley stated in his 1947 *Correspondence*.

6. "FINGERS IN THE PIE"

1. The word *Vergeltungs* may also carry the slightly different shading of "retaliation," which Hitler was especially eager for after British bombers introduced attacks on German cities in March 1942. Coined by Goebbels's Propaganda Ministry, this nomenclature was first applied publicly to the "V-1" jet-propelled flying bomb— which the Allies dubbed the "buzz bomb," a forerunner of today's cruise missiles— on June 16, 1942, the day after the premier volley was fired against London (the number "1" was meant to imply that additional types were in store). The "V" also deprecated Churchill's famous sign for victory (Robert Lusser to W. A. Heflin, 25 July 1958, in Wernher von Braun papers, box 3, Library of Congress).

2. Jean Mialet, *"Dora, le camp oublié,"* in *Bulletin de l'Amicale des Prisonniers Politiques de Dora*, 1995 (spring).

3. Hajo Holborn, *A History of Modern Germany, 1840–1945* (Princeton University Press, 1982), p. 801.

4. Walter Dornberger addressed the charge that the V-2 contributed to the defeat of Germany in an 1963 overview of the rocket program. See Walter Dornberger, "The German V-2," *Technology and Culture*, fall 1963, pp. 393–409.

5. "Affidavit of Membership in NSDAP of Prof. Dr. Wernher von Braun," June 18, 1947, El Paso, Texas. This secret deposition was taken during consideration of his U.S. immigration status under Project Paperclip. FOIA request.

6. Heinz Höhne, *The Order of the Death's Head* (Penguin Books, 2000), p. 137. The contradictions involved in signing up newcomers from the moneyed Wilhelmine ruling class were not lost on the old brown-shirted foot-soldiers, who called them "March Violets" (ibid., p. 135). Himmler liked to think of the SS as comparable to the medieval Teutonic Knights (Holborn, p. 749).

7. Alexandra Richie, *Faust's Metropolis* (Carroll & Graf, 1998), p. 425.

8. Ibid., p. 447.

9. It was in *The New York Times Magazine* of May 6, 1934, that Gertrude Stein made her sarcastic remark that "Hitler ought to have the [Nobel] peace prize, because he is removing all the elements of contest and of struggle from Germany. By driving out the Jews and the democratic and left element, he is driving out everything that conduces to activity."

10. Braun's "Affidavit of Membership." Some historians have argued that von Braun joined at this time for "recreation" and because Nazi student organizations were pressuring students who were not Party members to demonstrate their ideological conformity. See Michael Neufeld, "Wernher von Braun, the SS, and concentration camp labor: Questions of moral, political, and criminal responsibility," *German*

Studies Review 25, no. 1 (2002). If fresh air and exercise on horseback were his desire, there were of course many other opportunities besides those offered by the SS. Though the nature of his research for the army was secret, his status as one of Karl Becker's special graduate students would clearly have mitigated any peer pressure. In the "Affidavit," he stated that he "got [his] discharge" from the *Reitersturm* "in summer 1934," but did not explain what this entailed. On July 2, 1934, Anton Freiherr von Hohberg und Buchwald, the leading SS horseman in East Prussia, was shot by SS men for giving away secrets to the Reichswehr (Höhne, p. 138). It is not known whether von Braun left before or after this incident. In any case, he never commented on these matters beyond the few terse sentences cited here.

11. Von Braun's "Affidavit of Membership."

12. See Michael Neufeld, "Hitler, the V-2, and the battle for priority, 1939–1943," *Journal of Military History* 57 (July 1993), pp. 511–38. Military planners had posited a war starting sometime between 1942 and 1944, not 1939, and were thus faced with severe munitions shortages after the invasion of Poland as Hitler plunged ahead with campaigns against France and the Soviet Union.

13. Dornberger claimed that Becker "committed suicide because of a quarrel with Hitler" (Walter Dornberger, *V-2*, [Viking Press, 1954], p. 70).

14. Von Braun's "Affidavit of Membership." Promotions in SS membership file. FOIA request.

15. Albert Speer, *The Slave State* (Weidenfeld and Nicolson, 1981), p. 25.

16. Himmler was of course already responsible for SS atrocities in Poland, which he knew could send him to the gallows if Germany were defeated in the war.

17. Speer, *Slave State*, pp. 18 and 61. Of course, Speer was in a far more powerful position than von Braun to rebuff Himmler.

18. Dornberger, *V-2*, p. 47, and "The German V-2," p. 398. The Kaiser Wilhelm Geschütz, or "Paris Gun," was first fired in combat in March 1918 from a forest about 75 miles from Paris. A battery of three guns transported by railroad launched waist-high, 210-mm shells carrying about 20 pounds of TNT. The shells arced into the stratosphere, the first man-made objects to do so. (Dornberger envisioned a rocket that would send a hundred times more explosive for twice the range with far better accuracy.) Initially, Parisians were mystified about what was happening, since they could hear no airplanes or heavy artillery. Before the guns were returned to Germany and destroyed as the Allies advanced, more than 300 shells were fired, killing some 250 people. The deadliest incident occurred on March 29, when a shell collapsed the roof of Saint Gervais church, near the Hotel de Ville in central Paris, on Good Friday, killing 88 worshippers and wounding 68. Because of poor accuracy, rapid barrel wearout, and the relatively small amount of damage inflicted, the guns were of little military value, being of mostly psychological impact on the French and of propaganda value on the home front. They were banned by the Treaty of Versailles, though the Allies never captured one. See Gerald V. Bull and Charles H. Murphy, *Paris Kanonen—The Paris Guns* (Presidio Press, 1991).

19. Holborn, p. 745. About a million men were in the SA (*Sturmabteilung*), with

another 3.5 million in reserves. The SS (*Schutzstaffel*) arose out of Hitler's personal bodyguard and became virtually a parallel government by the end of World War II. The SD (*Sicherheitsdienst*) was the security service of the SS empire.

20. Von Braun was quoted many years later as saying that his impression of Hitler at this time was of "a fairly shabby fellow," though there is no way to know whether he was speaking about Hitler's character or attire (the führer as photographed was neatly dressed in leather greatcoat, polished boots, white shirt, and tie). In any case, he never recorded such disdain about anyone in the motley Raketenflugplatz crew. The quotation is from Bob Ward's *Wernher von Braun Anekdotisch*, a 1972 hagiographic collection of unreferenced remarks published in West Germany. Ward was a reporter and editor of the *Huntsville* (Alabama) *Times*.

21. Dornberger, *V-2*, p. 31.

22. Holborn, p. 723.

23. "Wenn ich mich selbst prüfe, ob ich seinerzeit mit all meinen alten Freunden und Kollegen zusammen im Hitler-kabinett geblieben wäre, so antworte ich—auch wenn ich damit meine mangelnde Voraussicht eingestehe—mit ja" (Magnus von Braun, *Von Ostpreussen bis Texas* [Stollham, 1955], p. 234). He referred to the criticism he received after the war from historians and journalists as "wisecracks," maintaining that no one in his position could have known what was to come. In general, the Wilhelmine élites' strategy of "taming" Hitler had nothing to do with defending the Republic against National Socialism, but was rather an attempt to gain complete control over a new authoritarian state of their own design. They were not directly responsible for every Nazi crime ahead, but they presided over the destruction of democracy and then threw their lot in with Hitler. As Detlev Peukert has written: "Hitler needed power, and the old élites needed mass support" (*The Weimar Republic* [Hill and Wang, 1992], p. 265). Von Papen himself told the Nuremburg court in 1945 that "there was no other way out" than to let Hitler into the government. Quoted in Richard Overy, *Interrogations: The Nazi Elite in Allied Hands, 1945* (Penguin Books, 2001), p. 172.

24. Magnus von Braun, p. 238. In a rare reference to his children, Magnus remembered fondly that "meiner drei Söhne" joined him and Gürtner in the Oberwiesenthal library for wine and conversation. Wernher never mentioned it.

25. German students could advance very quickly to the PhD by today's standards. Future Nobel Prize–winner Werner Heisenberg, for example, obtained his doctorate in physics from the University of Munich in 1923, at age twenty-two, three years after starting university education.

26. Wernher von Braun, "Reminiscences of German rocketry," *Journal of British Interplanetary Society* 15, no. 3 (May–June 1956). Contrary to von Braun's bleak description, Walter Dornberger recalled that "we were very proud of that test stand," which was "fully equipped with all available resources of measurement technique" (*V-2*, p. 23).

27. See Chapter 4, note 2, and Chapter 5, note 23.

28. Dornberger, *V-2*, p. 28.

29. Von Braun, "Reminiscences."
30. Copy in Wernher von Braun papers, Library of Congress, box 52.
31. Von Braun's principal academic adviser was Erich Schumann (grandson of the composer Robert Schumann), who held a doctorate in "systematic musicology" (acoustics) and taught courses on acoustics and explosives. He began working as a physicist for the German defense ministry in 1922 and joined the Nazi Party in 1933. Politics lay behind his faculty appointment at the University of Berlin. As a professor and military officer he played a role in forming science policy during the Third Reich—including the abortive nuclear energy program (though he did not thoroughly understand nuclear fission)—when he served as director of research for the *Heereswaffenamt* and scientific adviser to Field Marshall Wilhelm Keitel. In the company of scientists he wore his military uniform and saluted, but introduced himself in civilian attire as "Herr Professor" to generals. Leading physicists joked that he confused *Physik* with *Musik*. See Thomas Powers, *Heisenberg's War* (Knopf, 1993), pp. 130–31.
32. "Protokoll" of gestapo interview dated July 16, 1934; copy in Imperial War Museum (IWM) archive, file MI 14/801 (V).
33. Schneider to V2, dated December 22, 1933, in IWM MI 14/801 (V).
34. Zanssen to Abw, dated August 27, 1934, in IWM MI 14/801(V).
35. "Protokoll."
36. In a hero-worshipping biography published in Germany in 1969, at the apogee of von Braun's postwar fame, he was quoted as claiming that Max and Moritz "were entirely my own work," that "I designed them myself, I drew every screw on the drafting table," that "in short, I had put them together from A to Z." This biography and others like it were notoriously slanted, but if valid the quotation clearly contradicts Dornberger's detailed account of other vital participants in the program. See Bernd Ruland, *Wernher von Braun: Mein Leben für die Raumfahrt* (Burda [Offenburg], 1969). Burda was and is primarily a magazine publishing house, focused on fashion, part of the same company that since 1948 has sponsored the "Bambi" television and media celebrity prize.
37. Von Braun "Reminiscences," and Dornberger, *V-2*, p. 36.
38. Arthur Rudolph, "My rocket work in Germany," typescript copy. Also oral history interview of Rudolph by Michael Neufeld, Department of Space History, National Air and Space Museum, August 4, 1989. Rudolph (1906–1995) had been a member of the Nazi Party since 1931 and was also in the SA (from which the army made him resign when he became a Kummersdorf employee). Regarding organization at Kummersdorf, he recalled that "von Braun was to be in charge of the group in overall development, Riedel in charge of design, and I in charge of the workshops." He stated that von Braun's office was at Ordnance headquarters in Berlin with Dornberger and other military officers, and that von Braun spent about half his working time at Kummersdorf.
39. Dornberger, "The German V-2," p. 395.
40. Dornberger, *V-2*, p. 38. Dornberger was not present on Borkum, because he had

been transferred to train a battery of black-powder rocket launchers, which he had helped develop. As described by the U.S. Military Intelligence Service's "Tactical and Technical Trends" bulletins of World War II, these were eventually used to fire smoke and chemical agents on the Russian front and elsewhere (ibid., p. 36). This routine assignment would appear to indicate that the liquid-propelled rocket program was considered of peripheral importance at the time.

41. Ibid., p. 38.

42. Dornberger recalled persuading Commander-in-Chief of the Army Werner Freiherr von Fritsch to visit Kummersdorf in March 1936. "Hardly had the echo of the motors died away in the pine woods than the General assured us of his full support, provided we used the funds to turn our rocket drive into a serviceable weapon of war" (ibid., pp. 38–39). March 1936 also saw the German Army's remilitarization of the Rhineland, a flagrant violation of the Versailles and Locarno treaties, demonstrating France's diplomatic and military weakness.

43. Holborn, p. 766.

44. Ibid.

45. Von Braun, "Reminiscences." The great ace was of course Manfred von Richthofen, the Red Baron, a distant cousin. Wolfram also earned ace status in World War I and became a combat commander of the Condor Legion that bombed Guernica in 1937 during the Spanish Civil War. He eventually rose to be field marshall general in the Luftwaffe during World War II.

46. Ibid. Nearly 20 years after the fact, von Braun's admiration for the youthful verve of the Luftwaffe is apparent. It was a new branch of military service created by Hermann Göring and thoroughly Nazified, unlike the army and navy. Arthur Rudolph recalled a German saying of the time: "the Kaiser's Navy, the Prussian Army, and Hitler's Luftwaffe" (Rudolph interview; see note 38 above).

47. Ibid.

48. Wernher von Braun to Nazi Party official Theo Groneis, June 5, 1935, IWM MI 14/801(V).

49. Von Braun, "Reminiscences."

50. Rudolph interview; see note 38 above. Why von Braun would take along a shop foreman to such a lofty meeting cannot be determined from the record. It is fair to assume that Rudolph's longtime Nazi Party membership might have been of some advantage.

51. Dornberger remembered simply that "we succeeded in interesting" von Richthofen, by describing to him "in glowing terms the possibilities of using rocket motors as take-off help for heavy bombers and of equipping fighter aircraft with rocket power plants, and suggested building a combined [Army-Luftwaffe] establishment" (V-2, p. 39). In another text he mentioned these applications and added that "the big rocket was only a somewhat hazy dream" at this point ("The German V-2," p. 398). He said nothing about breaches of bureaucratic protocol.

52. Von Braun, "Reminiscences."

53. Dornberger, V-2, p. 39.

54. Ibid., pp. 38–39, 41. Kesselring (1885–1960) would go on to command German air units in the invasions of Poland, France, and the Soviet Union, as well as the Battle of Britain, and finally gain notoriety as the commander of forces in Italy. He was sent to the gallows for war crimes by the Nuremberg court, but the sentence was commuted to life imprisonment and he was released in 1952.

7. SUPREME ZEAL

1. Volkhard Bode and Gerhard Kaiser, *Raketenspuren, Peenemünde 1936–1996* (Bechtermünz Verlag, 1998), p. 24. Walter Dornberger, *V-2* (Viking Press, 1954), p. 40. Ahlbeck is today part of the municipality of Heringsdorf, adjacent to the Polish border, which comprises three old summer resorts known as the *Kaiserbäder* because they were once favored by the Wilhemine court. Along with less pretentious towns nearby, they have also been called the "bathtub of Berlin" because of their proximity for urban vacationers. Before the German military acquired Peenemünde and forced its residents to move, the relatively poor village had long tried, without success, to lure tourists from these nearby resorts. Axel Dietrich, *Peenemünde through the Centuries* (Axel Dietrich Verlag, Peenemünde, 1994). After languishing during the Soviet era, they have all been rejuvenated as popular beach destinations today, though Peenemünde remains physically scarred by the twentieth century.
2. These were Dornberger's words—"my beautiful Peenemünde"—upon seeing the damage after British bombers attacked the area on August 17, 1943. Quoted in Erik Bergaust, *Wernher von Braun* (National Space Institute, 1976), p. 32. In *V-2*, Dornberger remembered saying, "My poor, poor Peenemünde!" (p. 167).
3. The illustrious Haber, founding director of the institute in 1911, received the 1919 Nobel Prize in chemistry for the ammonia synthesis process. Though the institute was taken over by the Reichswehr during World War I, the Swedish Academy focused on man-made ammonia as a boon to world agriculture, rather than explosives, and also chose to ignore Haber's pivotal role in developing poisonous battlefield gases for the army, for which the Allies considered him a war criminal. Always a fervent nationalist, Haber nevertheless left Germany when the Nazis came to power in 1933, because he was Jewish. He died the following year in Switzerland. The institute is today named the Fritz Haber Institute in his honor.
4. Wernher von Braun, "Reminscences of German rocketry," *Journal of the British Interplanetary Society* 15, no. 3 (May–June 1956). It must be noted that the same argument about science's supposed separateness from society would be used to exonerate the Manhattan Project physicists.
5. SS-Stammrolle membership file, FOIA request. To his glider piloting in the early 1930s he added a license for powered flight in 1933. The German Sport Aviation Club (*Deutscher Luftsport-Verband*) that he joined while still a student became the National Socialist Aviation Corps (NSFK, or *Fliegerkorps*) in the summer of 1934, of which he remained a member until joining the Luftwaffe reserve ("Affidavit of Membership in NSDAP of Prof. Wernher von Braun," June 18, 1947, El Paso, TX.

FOIA request). The nationalistic context for learning how to fly in Germany during this period was expounded by Peter Fritzsche in *A Nation of Flyers: German Aviation and the Popular Imagination* (Harvard University Press, 1994). After 1933, the "swastika-emblazoned" *Deutscher Luftsport Verband* and the gliding movement, which National Socialists had always encouraged for its patriotic spirit, were heavily subsidized by the Nazis. See Peter Fritzsche, "Machine Dreams: Airmindedness and the Reinvention of Germany," *American Historical Review* 98, no. 3 (June 1993), p. 701.

6. Peter P. Wegener, *The Peenemünde Wind Tunnels, A Memoir* (Yale University Press, 1996), p. 16. Interestingly, Wegener "never heard a single remark about space flight" during his time at Peenemünde. "In my several meetings with von Braun, he never suggested this possibility, even in small social gatherings. In contrast, much of the postwar literature recounts the [V-2] development as if space flight had been a distant goal during the war" (ibid., p. 41).

7. Hitler was of course an avid amateur architect with a dictator's passion for building, but he was as riddled by contradictions in this field as in others. Though the National Socialists defamed Modernism in favor of a romanticized German past, structures associated with technology—airports, factories, autobahns, and bridges—were unabashedly Modernist. Luftwaffe buildings were especially known for their functionalist esthetic. Contrary to his reputation, Hitler was sometimes receptive to Bauhaus ideas, accepting Mies van der Rohe's designs for autobahn service stations, for example. Much state architecture reflected Hitler's version of an international style that can be seen today in federal buildings of the same era in Washington, DC, such as the National Archives on Pennsylvania Avenue. The Peenemünde buildings combined the steep gabled roofs of traditional structures in the region with the unadorned surfaces of Modernism. Thatched roofing, which is still common along the Baltic coast, was certainly not to be found at forward-looking Peenemünde, which featured sleek tiles. See Frederic Spotts, *Hitler and the Power of Aesthetics* (Overlook Press, 2002).

8. These are biographer Michael Neufeld's terms, but they faithfully echo the tone of postwar testimony by von Braun's contemporaries at Peenemünde (Michael J. Neufeld, *Von Braun: Dreamer of Space, Engineer of War* [Knopf, 2007], pp. 93–94). Von Braun was the chief recipient of such praise, but not the only one. Former Peenemünders referred to many of their colleagues as geniuses in the course of oral history interviews.

9. Wegener, p. 48. Wegener recalled that "von Braun's close associates had a tendency to flatter him, a fact that he himself did not seem to notice."

10. Dornberger to RAdm. D. S. Fahrney, USN (Ret.), November 21, 1961, Wernher von Braun papers, Library of Congress, box 6.

11. It also made him subject to ridicule in quarters that did not accept his reputation. The popular American cartoonist Al Capp drew a fatuous character called Dr. Werner von Brain, "the greatest mathematician on Earth," who came to the aid of star-struck generals. Syndicated cartoon, October 24, 1963.

12. Von Braun clearly had a powerful, charismatic effect on some men throughout his life. His friend Frederick C. Durant III, the postwar American spaceflight advocate and director of astronautics at the National Air and Space Museum, spoke of how "again and again, I have seen von Braun's personality work magic on opinionated individuals who had preconceived notions and erroneous impressions of von Braun himself, his projects and accomplishments," and of how "over and over I've watched these opinions change, usually within a few minutes of a first meeting, as von Braun's personal warmth, and engaging manner and obvious honesty are communicated as if by a sixth sense" (Bergaust, p. 18). Such remarks are remarkably similar to the salutary effects of meeting Hitler on British leaders such as David Lloyd George and George Lansbury, or the impressions of French ambassador François-Poncet: "[Hitler] excited curiosity; he awakened sympathy; his prestige grew; the force of attraction emanating from him had an impact beyond the borders of his country" (Ian Kershaw, *Hitler, 1936–1945: Nemesis* [W. W. Norton, 2001], p. 29). In the photographic collections contained in various heroic postwar biographies of von Braun, an uncanny number of portraits appear with no other apparent purpose than to display his physical beauty—calling to mind such photographs of John F. Kennedy. See for example E. Stuhlinger and F. Ordway, *Wernher von Braun, Crusader for Space* (Krieger Publishing Company, 1994), frontispiece.

13. A U.S. military security agent report on von Braun dated April 9, 1953, cited a knowledgeable informant who said that "when [von Braun] first began work in this country he was feared by the other technical men who were working with him." FBI files, FOIA request. The term "intellectual reparations" was used by advocates of exploiting German technical expertise after World War II. See John Gimbel, *Science, Technology and Reparations: Exploitation and Plunder in Postwar Germany* (Stanford University Press, 1990).

14. Kershaw, pp. xlii–xliii. Kershaw eloquently draws the haunting distinction between fanaticism and complicity.

15. Ibid., pp. 5–8 and 9–27.

16. Dornberger, p. 47.

17. Rudolph interview. "I can in fact not think of any instance where he [von Braun] told me pointblank, 'You change that,' " Rudolph said.

18. Dornberger, pp. 36, 47, and 49.

19. Von Braun, "Reminiscences," and Dornberger, p. 34. The German Navy had sought to compensate for Versailles Treaty limitations on the tonnage of its ships by improving their speed, navigation, and gun accuracy. See Donald MacKenzie, "The Soviet Union and Strategic Missile Guidance," *International Security* 13, no. 2 (Fall 1988), and *Inventing Accuracy* (MIT Press, 1990), chapter 2. Under another name until Hitler came to office, Kreiselgeräte had been owned secretly by the Navy since 1926, to develop self-contained inertial navigation systems. World War II German battleships such as the *Bismarck* and *Tirpitz* carried Kreiselgeräte gun barrel control systems to provide steady aim in rough seas.

20. Dornberger, p. 34. Boykow died in 1935, before he could see the project through to

completion. A former actor, he "had a somewhat broader concept of life in general than most of our engineers," von Braun remembered (von Braun, "Reminiscences.")

21. Oral history interview of Gerhard Reisig by Michael Neufeld, Department of Space History, National Air and Space Museum, June 5, 1985.

22. Dornberger, p. 53. Peenemünde's wind tunnel cavity would be four times larger, at 40-by-40 cm, than Aachen's.

23. Wegener, p. 44.

24. Von Braun, "Reminiscences." Only one of the Raketenflugplatz alumni is known to have left Germany after Hitler's rise to power: Willy Ley (1906–1969), who moved to Great Britain in 1935 and lived out his life in the United States as a science-fiction writer, spaceflight advocate, and apologist for the Peenemünders.

25. Dornberger, p. 50; von Braun, "Reminiscences."

26. Kershaw, p. 27.

27. He listed 7 Kirchstrasse as his address beginning in 1937 as part of a "Personal History Statement" signed on November 25, 1947, in conjunction with immigration to the United States. FOIA request.

28. See Michael Neufeld, "Rocket aircraft and the 'turbojet revolution': The Luftwaffe's quest for high-speed flight, 1935–39," in *Innovation and the Development of Flight*, Roger D. Launius, ed. (Texas A&M University Press, 1999).

29. See Roderick Stackelberg, *Hitler's Germany: Origins, Interpretations, Legacies* (Routledge, 1999), chapter 10, for a description of the labyrinthine rules aimed at banishing Jews from public life after 1933.

30. Letter dated January 22, 1971, in Bob Ward, *Dr. Space: The Life of Wernher von Braun* (Naval Institute Press, 2005), pp. 227–29. The letter, written while von Braun was working at NASA headquarters in Washington, DC, was a reply to an over-the-transom missive asking him why he had not used his influence during World War II to help the Jews. Though it was obvious that persecution existed before the war, von Braun answered, he had not been aware of any atrocities. He also wrote that Jews were welcome in the German Army until the war began. The Wright brothers first demonstrated their flying machine to the U.S. Army at Fort Myer, Virginia, in September 1908.

8. GRAND AND HORRIBLY WRONG

1. On October 15, 2008, a collection of his sketches of futuristic spaceships sold at auction in Los Angeles for $132,000.

2. This is the date for when he "entered" the Party found on numerous postwar U.S. government security reports that referenced NSDAP records in the Berlin Document Center. Those records indicate that he applied for membership (*Aufnahme beantragt am*) on November 11, 1937, membership then being retroactive to May 1, the Nazi *Tag der Arbeit*, a day of celebration of Party unity. Von Braun's SS membership file listed his *Eintritt in die Partei* as December 1, 1938. That would place it some three weeks after the Kristallnacht pogrom on November 9 and 10. These

various membership dates are significant inasmuch as they contradict his sworn U.S. Army "Affidavit," which made it appear that he did not join the Party until the war began.

3. "Affidavit of Membership in NSDAP of Prof. Wernher von Braun," June 18, 1947, El Paso, TX. FOIA request.

4. In retaliation for bombing the *Deutschland* (which had been constructed under Versailles treaty limitations on size, hence the British term "pocket"), a sister ship fired on the Spanish coastal town of Almería, killing 21 civilians.

5. Axel Dietrich, *Peenemünde through the Centuries* (Axel Dietrich Verlag).

6. Walter Dornberger, *V-2* (Viking Press, 1954), pp. 42–45; Wernher von Braun, "Reminiscences of German rocketry," *Journal of the British Interplanetary Society* 15, no. 3 (May–June 1956).

7. This must have been the first time the dye was used, because the seaside air would always have been humid enough to produce copious amounts of condensation on the cold rocket body.

8. Dornberger; von Braun, "Reminiscences."

9. Reisig oral history interview, National Air and Space Museum, June 5, 1989.

10. Von Braun, "Reminiscences."

11. Ibid.

12. Ian Kershaw, *Hitler, 1936–1945: Nemesis* (W. W. Norton, 2001), pp. 46–51, and Holborn, pp. 773–74.

13. Kershaw, pp. 52–58, and Holborn, pp. 774–75. Other historians have considered the possibility that Blomberg and Fritsch were framed by Göring and Himmler. See Peter Padfield, *Himmler* (Henry Holt and Company, 1990), pp. 212–13. Padfield supplies different details for the scandals, such as that Blomberg, "a handsome widower and known ladies' man," was introduced to his future bride by a hotel manager in Thuringia, and that the bachelor Fritsch had once "befriended two Hitler Youths and had been in the habit of instructing them in history and geography, punishing inattention by striking their bare calves with a ruler, a fetish which [SS] agents had picked up."

14. Kershaw, pp. 46–51. Fritsch was exonerated (but not rehabilitated in rank) in 1938 by a military court and mortally wounded during the invasion of Poland in September 1939. Blomberg died in prison in Nuremberg in 1946.

15. By this time, of course, the army used its independence mostly to ward off bureaucratic subservience, since it had been politically aligned with the far right throughout the Weimar era.

16. Quoted in Kershaw, p. 60. Perhaps this was what von Braun had in mind when he said that he was "officially demanded" to join the Party.

17. Rudolph oral history interview, National Air and Space Museum, August 4, 1989.

18. "Memo for Record, Subject: Dr. Paul Schroeder" dated July 29, 1958, and classified "Confidential," signed by Gordon L. Harris, a U.S. Army public information officer, in response to a query from journalists Jack Anderson and Drew Pearson regarding a 40-paged manuscript in their possession written by Schröder "which

contained uncomplimentary references" to von Braun. The memo related how Harris asked a number of prominent Peenemünde alumni, including Dornberger, about Schröder's "bitter attack," all of whom dismissed him as a "neurotic." Harris recommended that Schröder, who was apparently applying for U.S. citizenship and a job at the Jet Propulsion Laboratory in California, not be employed "by any element of the Command." Schröder had not been among the core group of rocket technologists who came with von Braun to the United States.

19. Erik Bergaust, *Wernher von Braun* (National Space Institute, 1976), p. 21.

20. Rudolph oral history.

21. Dornberger, pp. 139–40.

22. Gerhard Reisig oral history interview, National Air and Space Museum, June 27, 1985.

23. Holborn, p. 777.

24. Kershaw, p. 75.

25. Kershaw, pp. 108–25, and Holborn, pp. 777–87. For a contemporary description of preparations for war in London, see the letter from Graham Greene to his mother dated September 27, 1938, in Richard Greene, ed., *Graham Greene, A Life in Letters* (W. W. Norton, 2008), p. 93.

26. Dornberger, pp. 64–66.

9. DEPRAVITY

1. "Himmler's prestige was represented solely by the power concentrated in him," Albert Speer remembered. "If he had been deprived of his control over life and death, over spying by the Gestapo and the Security Service, then he would have been deprived of power overnight." Speer called him "colorless," "inconspicuous," and "an utterly insignificant personality who, in some inexplicable manner, had risen to a high position" (Albert Speer, *The Slave State* [Weidenfeld & Nicolson, 1981], pp. 27–28). Dornberger wrote that he looked like "an intelligent elementary school teacher" (Walter Dornberger, *V-2* [Viking Press, 1954] p. 180).

2. Reisig oral history interview, National Air and Space Museum, June 5, 1989. Von Braun estimated that 60–65,000 drawing modifications were required before the experimental A-4 reached the production stage. "Survey of the development of liquid rockets in Germany and their future prospects (by Professor von Braun)" in *Report on Certain Phases of War Research in Germany*, vol. 1, prepared by F. Zwicky, Aerojet Engineering Corporation, October 1, 1945. The production target was five hundred missiles per year around the beginning of 1940—an essentially pointless quantity from a strategic standpoint even if it had been quickly attainable. Eventually it was set at 5000 per year. An official production order dated October 19, 1943, was for 12,000 missiles at a rate of 900 per month costing 40,000 marks apiece. Building factories for an untested rocket seems to have bothered only Hitler.

3. Dornberger, p. 232. See chapter 25 for his justification of the way Peenemünde was organized.

4. Deposition of Wernher von Braun before the German Consulate General, New Orleans, February 7, 1969.

5. British Prime Minister Winston Churchill was an early advocate of summary execution of German leaders and remained so until the end of the war, though the other Allies remained intent on court trials. See "Churchill: Execute Hitler without trial" in *The* (London) *Sunday Times*, January 1, 2006. President Roosevelt supported unconditional surrender at Casablanca, while Churchill and Stalin argued that demanding it would prolong the war.

6. From a regime for which photographic record was an obsession, there is only one extant picture of von Braun wearing his SS uniform, during a visit by Himmler to Peenemünde in June 1943, wherein von Braun's face is partially obscured behind the Reichsführer SS. The photo was first published after the war by Dornberger (in *V-2*), who surely knew what he was doing. There is reliable testimony by numerous individuals who saw von Braun wearing the black garb, including an instance in May 1942 at one of the Peenemünde test stands, the kind of event that would have been copiously photographed. It is reasonable to conclude that many photos of von Braun in SS uniform were purposely destroyed to protect his reputation. Perhaps some still exist (Volkhard Bode and Gerhard Kaiser, *Raketenspuren* [Bechtermünz Verlag, 1998], p. 46).

7. Speer, *Slave State*, p. 18.

8. Regarding the deterioration of the army-Luftwaffe partnership at Peenemünde, Gerhard Reisig maintained that the "politically smart" SS said, in effect: " 'Well, you are fighting each other, we have to do it for our country and we will take it over.' And I think they were right." The SS, he added, "improved the efficiency of the whole operation" (ibid.).

9. Albert Speer, *Inside the Third Reich* (Macmillan Company, 1970), p. 366.

10. Speer, *Slave State*, pp. 203–4. Dornberger's request went first to Lieutenant Colonel Gerhard Stegmaier, Peenemünde's commandant and a Himmler sycophant, who passed it on to Berger. Speer saw this as an attempt to skirt around himself and Army Armaments Chief General Friedrich Fromm, who was "still hesitant" about the rocket program. "Thus, direct lines were set up from a competent army officer to involve Himmler's authority directly." In January 1943, Dornberger used the Stegmaier-Berger-Himmler pipeline again to seek from Hitler a rise of the rocket program's priority for electrical equipment. "In this way, it was easy for the SS to infiltrate," Speer observed.

11. "In Peenemünde they are building a paradise," wrote first Armaments Minister Fritz Todt in July 1941 to General Friedrich Fromm, the Chief of Army Armaments who had approved a blank-check plan generated by Dornberger in the spring of 1940 (later rejected by Commander-in-Chief von Brauchitsch) that would have given Peenemünde's rockets the highest priority in the entire Reich arsenal. Todt urged the army to build in a manner appropriate to war. He died in an airplane crash in February 1942. (Quoted in Michael Neufeld, "Hitler, the V-2, and the Battle for Priority, 1939–1943," *Journal of Military History* 57 [July 1993], p. 527.)

12. Speer, *Inside*, pp. 365–66. Von Braun estimated the construction cost of Peene-münde at 300 million marks in "Summary of Development of Liquid Rockets in Germany" (ibid.). Total investment in the rocket program was in the billions.

13. Dornberger, *V-2*, pp. 141–43. Neufeld, p. 139.

14. *V-2*, p. 70.

15. Speer, *Inside*, p. 366.

16. For a landmark study of his motivations and evasions, see Gitta Sereny, *Albert Speer: His Battle with Truth* (Knopf, 1995).

17. Speer, *Slave State*, p. 18.

18. Ibid., p. 40.

19. I am grateful to the association of camp survivors, Amicale des Déportés Politiques et de la Résistance de Dora-Ellrich, Harzungen et Kommandos Annexes, for extensive literature written by members about their experience.

20. In April 1943, Rudolph traveled to the Heinkel-Flugzeug-Werke aircraft plant in Oranienberg, near central Berlin, to learn about the SS exploitation of prisoners from Sachsenhausen. He enthusiastically approved of what he saw and recom-mended such a system for Peenemünde. Bode and Kaiser, p. 54. (Slave labor was also used at two other missile production lines, established at the Zeppelin hangars in Friedrichshafen and the old Raxwerke locomotive plant 30 miles south of Vienna, Austria. The goal was to produce 900 units per month at the three loca-tions, but they were all bombed by the Allies.) Rudolph's ability to dissemble about his wartime experience was exhibited in a 1985 statement claiming he had learned that forced labor would be used at Mittelbau-Dora "to my horror" and that "I had the overpowering awful feeling that I was trapped in a cage like an animal, and I was trapped, as I experienced more and more, in the claws of the SS" ("My rocket work in Germany," typescript dated May 30, 1985).

21. A German reference published in 1995 states that Dornberger referred to the labor-ers at Peenemünde as "Morder, Diebe, Verbrecher" (murderers, thieves, criminals). This characterization would have been in keeping with longstanding Nazi propa-ganda about the camps. See K. Friedrich Baudrexl in Torsten Hess, Thomas A. Seidel, et al., *Vernichtung durch Fortschritt: Am Bespiel der Raketenproduktion im Konzentrationslager Mittelbau* (Westkreuz, 1995).

22. I am grateful to former Dora prisoner George Soubirou for this reference and for providing Sadron's name and academic affiliation to me in 1995.

23. The best account of the RAF operation remains Martin Middlebrook's *The Peene-münde Raid* (Allen Lane, 1982). The raid was specifically intended to eliminate Peenemünde's braintrust, but the only crucial loss was engine designer Walter Thiel.

24. Even though the Peenemünde technicians had succeeded in getting the A-4 off the ground, Speer remembered, Hitler had the "gravest doubts" and wondered whether a guidance system could ever be developed (Speer, *Inside*, p. 367).

25. According to Speer, it was this test that convinced him that "they could start manufacturing the A-4 despite all risks" (Speer, *Slave State*, p. 204). The launch

was part of a shoot-off competition on May 26 between the army's A-4 and the Luftwaffe's "buzz bomb" cruise missile, code-named Cherrystone, that would come to be famously known as the V-1. As a stellar audience of Reich leaders looked on from the Baltic coast at Peenemünde, one A-4 performed up to expectations, another somewhat less so, and two Cherrystones crashed into the nearby sea. Both programs were approved for mass production and Dornberger was promoted to Major General (Dornberger, *V-2*, pp. 95–98).

26. Documents obtained through FOIA request, but they have been available at the National Archives since the 1980s. Various researchers have attempted to locate Brill over the years, without success. One's imagination shudders at the possibility that by calling Himmler's attention to Brill, von Braun made her vulnerable to extermination. A monument to 1723 Steglitz Jews deported to concentration camps between 1941 and 1945, called the *Spiegelwand* for its mirrored surface that reflects passersby, was completed at the site of the former synagogue there—destroyed by Allied bombing in 1943—in 1995. Of some 3000 Jews living in Steglitz during the 1930s, out of an overall population of 200,000, about 2000 were left by 1939 due to emigration, of which 145 survived in 1945 (Donald W. Shriver, *Honest Patriots* [Oxford University Press, 2005], p. 42). Biographer Michael Neufeld takes seriously a bodice-ripper letter sent to von Braun at NASA in 1963 from a Frenchwoman claiming to have been his mistress in Paris in 1943, but though von Braun took the time to translate it himself, he did not answer it, thus relegating it to the category of crackpot mail (of which he received piles). One must doubt that he would have kept a potential blackmail letter in his office files, which in the two extant collections seem highly scrubbed (Michael J. Neufeld, *Von Braun: Dreamer of Space, Engineer of War* [Knopf, 2007], pp. 147–48).

27. Dornberger devoted a whole chapter (19) to this visit in *V-2*. In the previous chapter he placed Himmler's "unexpected" first visit at the beginning of April 1943, off by four months, in an otherwise detailed account devoted to showing how Himmler's offer to "employ his powers on our behalf" was rebuffed.

28. Speer, *Inside,* p. 368.

29. Kershaw, *Hitler 1936–1945*, p. 612.

30. Dornberger, *V-2*, pp. 100–106.

31. Speer, *Slave State*, p. 207.

32. Ibid., pp. 137 and 204. Emphasis in original.

33. Speer, *Inside*, p. 368. If the quotation is true, then it offers peculiar evidence that the führer had a conscience, however small.

34. Dornberger, *V-2*, pp. 198–99 and Speer, *Slave State*, p. 12.

35. All Dora prisoner numbers and the narrative chronology are from the encyclopedic *Produktion des Todes Das KZ Mittelbau-Dora* by Jens-Christian Wagner (Wallstein Verlag, 2001). On January 22, 1944, 44 men out of the 1000 who arrived that day from Buchenwald were identified as homosexuals. On January 27, 396 men were transported some 230 kilometers from Sachsenhausen-Oranienburg north of Berlin. On May 26, the first shipment of 200 men identified as Jews arrived from

Buchenwald, then 200 more on May 27, 28, and 29, and 120 on May 30. In January 1945, as the Red Army advanced through Poland and Himmler ordered the gas chambers there demolished and remaining prisoners executed, hundreds of prisoners (some shipments entirely of Jewish women) began to be transported to Dora from Auschwitz, more than 400 kilometers away.

36. Yves Béon, *Planet Dora* (Westview Press, 1997), p. 3. I am grateful to Mr. Béon for his generosity in describing his experience as a Dora prisoner.

37. "In early December 1944, Dr. A. Poschmann, chief physician of the Todt Organization, told me that he had seen Dante's Inferno" (Speer, *Slave State*, p. 210). This was not Speer's impression as presented to the Nuremberg court, however, where he said that conditions were "about the same as those on a night shift in a regular industry."

38. Researchers have sometimes had no choice but to piece together von Braun's whereabouts during the Third Reich era by examining his airplane pilot logbook, one of the few pre-1945 personal documents to find its way into collections of his papers in the Library of Congress and the U.S. Space and Rocket Center—home of "Space Camp"—in Huntsville, Alabama (which holds the logbook). Some years are missing. How and why he salvaged this artifact, while losing virtually everything else, is one of many peculiarities facing historians. I cite dates mined by Michael Neufeld in various publications.

39. Von Braun's official affiliation was always Peenemünde, never Mittelwerk, reflecting the nominal separation between development and production, which deeply overlapped in the workaday world. An organizational roster of 28 unalphabetized names attending a meeting on May 6, 1944, listed "Prof. v. Braun" fourteenth as working for "HAP 11," a code name for Peenemünde adopted in May 1943 that stood for *Heimat Artillerie Park 11* (Home Artillery Grounds 11). Dornberger was listed at the top as *B.z.b.V Heer*, for *Beauftragter zur besonderen Verwendung Heer* (Army Commissioner for Special Tasks), which meant that he was no longer directly involved with Peenemünde-East. Next was Major General Josef Rossmann in *Wa A Wa Prüf 1o* (an acronym for liquid-fuel rockets, as opposed to solid-fuel devices). The last ten names were affiliated with *Mittelwerk G.m.b.H.* (Central Works Ltd.) and included SS-Sturmbannführer (Major) Otto Forschner, the Dora camp commandant, and Direktor Arthur Rudolph, chief of production. This tortuous structure was largely beside the point as the SS's power rose, but it did mean that von Braun and Dornberger were less closely connected in the chain of command. Reflecting the fantasy that the missile factory would continue in peacetime as a quasi-governmental commercial company, presumably for the profit of the SS, legal formalities (such as the *G.m.b.H.* suffix) were put in place for production contracts and facility rents between the army and Mittelwerk—all for naught, of course.

40. *Amicale.*

41. Speer, *Inside*, pp. 370–71. The quotation obviously refers not to Speer and his party, but to on-site Mittelwerk managers.

42. Joachim Fest, *Speer* (Harcourt, 2001), p. 179.
43. Speer, *Slave State*, p. 211.
44. Ibid., p. 219.
45. See Matthias Schmidt, *Albert Speer: The End of a Myth* (St. Martin's Press, 1984).
46. "At times it seems as though, even in retrospect, Speer was hardly aware of the compartmentalization of his own thinking" (Fest, p. 181).
47. "Sworn Statement of Wernher von Braun, 14 October 1947, Fort Bliss, Texas, USA" (text in German only). Rickhey, the highest Mittelbau-Dora official tried before the Nuremberg court, was acquitted (prosecutors focused on mistreatment of individual prisoners) and the trial record was classified for the next three decades. The relevant numbered questions and von Braun's responses were as follows:

> 10. Haben Sie selbst in der Fabrik Mittelwerke in Nordhausen gearbeitet? Wenn ja, geben Sie bitte Daten an.
> Nein.
> 11. Wenn Sie nicht in den Mittelwerken beschaeftigt waren, besuchenten Sie jemals die Fabrik? Wenn ja, wann und in welcher Eigenschaft?
> Ja. Ich war zum ersten Male etwa September odor Oktober 1943 in dem spaeteren Mittelwerk, als die Tunnel noch das Oellager der Wifo beherbergten. Spaeter, als die A-4 Fertigung in Gang gekommen war, war ich noch etwa 15 bis 20 mal dort, stets um irgendwelche technishen Fragen zu besprechen, die mit technischen Aenderungen am A-4 zusammenhingen. Das letzte Mal war ich etwa Februar 1945 dort.
> 12. Haben Sie waehrend Ihres Besuches die allgemeinen Arbeitsbedingungen in der Fabrik Mittelwerke zu irgendeiner Zeit von Mai 1944 bis April 1945 beobachtet?
> Die Arbeitsbedingungen im Mittelwerk haben sich waehrend der gesamten Zeit von ende 1943 bis zu meinem letzten Besuch laufend gebessert. Sie waren in der ersten Zeit aeusserst primitiv, da die nur fuer Oellagerung bestimmten Tunnel in keiner Weise fuer die Aufnahme einer akkuraten Fertigung und fuer die Aufnahme von vielen tausend Menschen bereit waren. Da noch kein Lager vorhanden war, wohnten die Gefangenen unter primitisten Bedingungen in den Tunnels selbst. Ab sommer 1944 waren erhebliche Verbesserungen vorhanden oder teilweise auch noch in Bau:
> — Krananlagen erleichterten den Gefangenen den Umgang mit schweren Teilen.
> — Sanitaere Anlagen verbesserten die hygienischen Verhaeltnisse.
> — Die Luftverhaeltnisse waren in Ordnung gekommen.
> — Die Beleuchtung war auf einen Stand gebract worden, der saubere Arbeit ermoeglichte und die Augen schonte.
> Ein Lager fuer die Gefangenen war ausserhalb des Tunnelsystems erstellet worden.

48. Deposition to the German Consulate General, New Orleans, Louisiana, in the trial of Helmut Bischoff, Erwin Busta, and Ernst Sander (text in German only). Bischoff was a Mittelbau security chief, Busta an SS guard, and Sander a gestapo officer. Bischoff was released before sentencing for health reasons. Busta and Sander were sentenced to prison terms of 7.5 and 8.5 years, respectively.

49. The best known was perhaps Jean Michel's *Dora*, whose first American edition appeared in 1980.

50. I am grateful to the late Guido Zembsch-Schreve, Dora prisoner no. 77249, for providing the translated texts of these affidavits to me in 1997.

51. Guy Morand, "Testimony concerning the Nazi von Braun," dated September 8, 1995.

52. Robert Cazabonne, "Testimony concerning Dora concentration camp," dated February 3, 1997.

53. I am grateful to Mr. Jouanin for describing this incident to me (with a hand-drawn diagram of where he stood in the missile) during the April 1995 reunion of former prisoners at Dora to mark the fiftieth anniversary of the camp's liberation.

54. Von Braun to A. Sawatzki, August 15, 1944. In this letter—which would appear to have placed him in violation of Nuremberg war crime standards—he mentioned being introduced to a prisoner who was a French physics professor, now known to be Charles Sadron of the University of Strasbourg. He asked that Dora's camp commandant, SS Major Otto Förschner, be approached about letting the professor wear civilian clothes instead of the striped prisoner's uniform "so that his morale might improve to a point of eliciting real collaboration." Sadron refused to be helped. The minutes of a large meeting held to inaugurate Georg Rickhey as Mittelwerk's general director on May 6, 1944 (the list described here in note 39 recorded the 28 attendees, including von Braun) showed Sawatzki requesting 1800 more slaves from Hans Kammler and that the use of prisoners from Buchenwald was discussed among other routine production matters.

55. Thomas Franklin, *An American in Exile: The Story of Arthur Rudolph* (Huntsville, AL: Christopher Kaylor Company, 1987), pp. 78–79.

10. "A PSYCHOLOGICAL BLOCK"

1. In 2008 dollars.

2. See Wayne Biddle, "A great new enterprise," *Air & Space Smithsonian*, June/July 1989, pp. 30–38.

3. Stanley Kubrick's 1964 *Dr. Strangelove* (last line: "Mein Führer, I can walk!") and Tom Lehrer's "Wernher von Braun" on the 1965 album *That Was the Year That Was* (" 'Once the rockets are up, who cares where they come down? That's not my department,' says Wernher von Braun."). Norman Mailer's 1970 book, *Of a Fire on the Moon*, expressed the widespread Jewish skepticism about von Braun ("Still what a grip he had on the jugular of the closet missionary in every Wasp").

4. Dispatching his brother among slave laborers raises issues of ignorance vs. inten-

tion. Some of the allegations of mistreatment made after the war by former Dora prisoners might have involved either of the von Brauns. One conclusion is that since Wernher found no reason to forego being in the tunnels himself, he saw no reason to keep Magnus out. It might have been a purely technocratic matter for them, or they might have seen nothing wrong with the entire situation. There is no contemporaneous record of what they thought. Sending Magnus down the mountain alone to meet the U.S. Army has often been explained by the fact that he spoke some English, but Wernher also knew the language well enough to become the group's spokesman after their surrender. (Wernher could not have ridden a bicycle, because his broken arm was in an awkward cast.) In a prominent interview after the war, Wernher gave the date as May 10, which would have placed it after the official German capitulation to the Allies on May 7, thus absolving them of treason (Daniel Lang, "A Romantic Urge," *The New Yorker*, April 21, 1951, p. 87).

5. Details about the encounter from an interview with Schneikert in the *Herald Times Reporter* (Manitowac-Two Rivers, WI), August 22, 1977, and from "Mission Accomplished," the 44th Infantry Division battle history. Besides Wernher and Magnus von Braun and Dornberger, the initial group included Dieter Huzel and Bernhard Tessmann, two Peenemünde engineers who knew where a cache of some 14 tons of rocket program documents had been buried in a mine shaft in the Harz region under Wernher von Braun's orders. A U.S. Army security investigation report on Wernher von Braun dated July 2, 1953, noted that he "at one time, was not fully cooperating with American authorities in that he apparently intended to use the location of certain hidden scientific documents as a bargaining lever with U.S. officials" (Author's collection via FOIA).

6. For complete lyrics see John M. Coggeshall, " 'One of those intangibles': The manifestation of ethnic identity in southwestern Illinois," *The Journal of American Folklore* 99, no. 392 (April–June 1986), pp. 197–98.

7. David Johnson, *V-1, V-2: Hitler's Vengeance on London* (Stein & Day, 1981), pp. 194–95. One of the last V-2 attacks, on March 27, 1945, destroyed a block of flats in Stepney, East London, killing 134 and injuring 49. A total of 1054 rockets struck throughout southern England. Out of about 6000 V-2s produced in all, some 3225 were fired in anger—more at targets on the Continent, such as Antwerp (1610), than at England—and killed about 5000 people. Total output at Mittelbau-Dora was nearly 5800, showing that many turned out to be useless as weapons. As a point of comparison, the Allied fire-bombing of Hamburg in July 1943, which fueled Hitler's desire for vengeance, left some 50,000 dead.

8. "German Scientists," letter dated May 20, 1945, signed by Major General Clayton Bissell, recommending that "approval be given to a project of bringing to the United States a number of German scientists for exploitation by the technical services with a view to increasing our war making capacity against Japan." I am grateful to Kai Bird for providing this and other documents from the National Archives related to the transfer of Germans to the United States.

9. Lang, p. 87. Earlier in the article, Maj. James P. Hamill, a thirty-one-year-old Army

administrator of the German rocket group, was quoted as saying that "Peenemünde was Germany's Oak Ridge." Because all of von Braun's public interviews and writings were closely supervised by the army at this time, it is reasonable to assume that his comparison of the V-2 program and the Manhattan Project was an approved trope.

10. Lang, p. 86.

11. Von Braun's power to decide the fate of these men certainly helped create the fear of his authority, noted in postwar security reports, among those Germans who came with him to the United States (CIC report by agent Milton F. Gidge, dated April 9, 1953, in author's FOIA collection).

12. See Clarence G. Lasby, *Project Paperclip* (Atheneum, 1971), pp. 30–43. Though work in the Mittelbau tunnels was by this time pointless and essentially defunct, several important engineers stayed behind in makeshift offices at Bleicherode, a village about 10 miles from Nordhausen, including von Braun's deputy, Eberhard Rees, and the design chief, Walther Riedel. "We just continued, as a soldier didn't walk away," Rees said many years later (Interview of Eberhard Rees by Neufeld, November 8, 1989, p. 41; Department of Space History, National Air & Space Museum). Riedel was arrested by American CIC agents who mistook him for someone they suspected of working on chemical and biological weapons. His teeth were broken during interrogation, a fact that perhaps explains why von Braun later remarked to a reporter for *The New Yorker* that "when we reached the CIC, I wasn't kicked in the teeth or anything" (James McGovern, *Crossbow and Overcast* [Hutchinson, 1965], p. 166; Lang, p. 87).

13. Of about 5 million prisoners of war, 2 million died in German captivity, with another million unaccounted for. About 5 million foreign workers were obtained (at most 200,000 voluntarily), of which close to 2 million came from Russia and 750,000 from France. Indifferent care produced tragic mortality rates (Hajo Holborn, *A History of Modern Germany, 1840–1945* [Princeton University Press, 1982], pp. 809–10).

14. Lang, p. 87. Asked why he wanted to join the Americans, he replied with a strain of pure opportunism: "My country had lost two wars in my young lifetime. The next time, I wanted to be on the winning side."

15. Walter Dornberger, *V-2* (Viking Press, 1954), p. 271. Albert Speer recalled his last meeting "in early April" with Kammler, who "was planning to contact the Americans." "In exchange for their guarantee of his freedom, he would offer them the entire technology of our jet planes, as well as the A-4 rocket and other important developments, including the transcontinental rocket [A-10]. For this purpose, he was assembling all development experts in Upper Bavaria in order to hand them over to the Americans" (Albert Speer, *The Slave State* [Weidenfeld & Nicolson, 1981], p. 243). Perhaps some important engineers stayed in Thuringia to avoid being part of this ludicrous plan.

16. *Historique de l'Amicale des Déportés,* 1989; *Dora, Bulletin trimestriel de l'Amicale des Prisonniers Politiques,* Spring 1995.

17. Interview of Werner Karl Dahm by Neufeld, January 25, 1990, p. 11; Department of Space History, National Air & Space Museum. Looking busy was not just good for careerism. On October 18, 1944, Hitler ordered the formation of the *Volkssturm*, a people's militia wherein all able-bodied men between 16 and 60 were expected to bear arms.

18. If not Verne, then perhaps Oberth, whom Dornberger and von Braun had harbored at Peenemünde in 1941–43. He proved to be a fish-out-of-water in the military setting, but did produce a study of multistage rockets. Speer recalled that on October 20, 1943—one day after a written order was issued for mass production of 12,000 A-4s—an agreement was reached between the army and the SS to establish an underground test site near Traunstein, in southeastern Bavaria, for the development of an "American rocket" with ten times the A-4's thrust (Speer, *Slave State*, p. 210). The project was never fulfilled.

19. Dahm interview, p. 8.

20. There is no record that the Germans had any knowledge of the Manhattan Project, the secret American program to build an atomic bomb. But Speer recalled that on June 22, 1944, Himmler stated that "explosives are suddenly emerging whose speed and effect overshadow the newest explosives of our retailiatory weapons," meaning atomic weapons, and rebuked Speer for neglecting research (Speer, *Slave State*, p. 150). Rumors that the United States was threatening to destroy Dresden with an atomic bomb if Germany did not surrender before August 1944 circulated widely that summer in Berlin scientific circles after the German embassy in Lisbon filed such a report. See Thomas Powers, *Heisenberg's War* (Alfred A. Knopf, 1993), pp. 348–49. German atomic bomb research during the war never went beyond elementary theoretical studies. Von Braun told U.S. Army interrogator Frank Zwicky in May 1945 that Werner Heisenberg, the Nobel physicist, "worked on the chain reaction problem for U_{235}" in Berlin-Dahlem (the Kaiser Wilhelm Institute for Physics, opened in 1938 with funding from the Rockefeller Foundation, but taken over by the German Army in October 1939 for uranium research) and that he personally "saw Heisenberg in 1942 the last time." Frank Zwicky, "Report on Certain Phases of War Research in Germany," October 1, 1945, Aerojet Engineering Corporation. (Zwicky was a professor of astrophysics at the California Institute of Technology.) There is no record of the Heisenberg–von Braun meetings, but it is reasonable to assume that von Braun knew something about nuclear energy and was intrigued by its potential for rocket propulsion, at least. In Walter Dornberger's statement for Zwicky, he wrote that "great interstellar distances can only be conquered through the use of nuclear reactions for propulsion, for which the basic prerequisites are in the development stage." Evidently Dornberger, too, had some "brain bubbles."

21. Lang, p. 87.

22. Ian Kershaw, *Hitler 1936–1945: Nemesis* (W. W. Norton, 2001), p. 838. For an insightful discussion of this phenomenon, see Richard Overy, *Interrogations* (Penguin Books, 2002), pp. 159–73. "The common distinctions between right and

wrong, apparently so simple, proved for many and complex reasons to be beyond the grasp of most of those interrogated," Overy found.

23. National Archives, RG260, OMGUS/FIAT. Walter Jessel died on April 11, 2008, at the age of 95 in Boulder, Colorado, where he was a leader of numerous environmental organizations. Obituary in Boulder Daily Camera, April 15, 2008.

24. Lang, pp. 85–86.

25. National Archives, G-2 Paperclip file, RG 319, via Kai Bird. (Declassified in July 1974.)

26. Quoted in Lasby, p. 38.

27. James McGovern, *Crossbow and Overcast* (Hutchinson, 1965), p. 121. Linda Hunt, *Secret Agenda* (St. Martin's Press, 1991), pp. 24–25.

28. Qualification sheet via FOIA. "Allgemeine," meaning general or universal, distinguished his membership from the "Waffen," meaning armed or weapons, SS. It is impossible to know when the word "Allgemeine" was typed in above a caret mark between "Entered" and "SS" on the form, though the keyface appears uniform.

29. "Basic Personnel Record," Military Intelligence Service, CPM Field Installation, Boston, MA, September 26, 1945. His age is listed incorrectly as 34. (Declassified in November 1984.) Author's files via FOIA.

30. See also Joachim Fest, *Speer* (Harcourt, 2001) p. 177.

31. Albert Speer, *Inside the Third Reich* (Macmillan Company, 1970), pp. 371–72.

32. Von Braun, "Reminiscences of German rocketry," *Journal of the British Interplanetary Society* 15, no. 3 (May–June 1956), section subtitled "Politics and Rocketry"; and Walter Dornberger, *V-2* (Viking Press, 1954), chapter 21.

33. In von Braun's 1947 "Affidavit," he stated that his meeting with Himmler was "in summer 1944" and that the arrest came "approximately eight weeks later." Von Braun was only in his mid-thirties and not suffering from any known physical trauma that might have affected his memory, so it is odd that he would have been so far off in dating such recent events, though his inability to remember well is not surprising psychologically. In the 1951 *New Yorker* article, he placed the meeting in "February, 1944" and the arrest "three weeks later." He repeated the February 1944 timeframe in his 1956 "Reminiscences," but said only that the arrest came sometime later at "2 o'clock one morning." Various biographers have written that he was in prison on his thirty-second birthday, March 23. The meeting date, for which there is no documentary evidence, can be reckoned only by triangulating von Braun's and Himmler's possible whereabouts—a fascinating game for von Braun buffs. Of lasting significance is the fogginess of the whole story and the purposes it served.

34. Fest, pp. 204–5, and Speer, *Slave State*, pp. 107 and 213. According to Speer, Hitler promised him on May 13, 1944, that "in the B. [Braun] matter so long as he is indispensable to me, he will be exempt from any prosecution, difficult as the general consequences might be." There is ample evidence that Gebhardt was trying to kill Speer. See *Slave State*, pp. 224–27. Gebhardt, who saw himself as a "political doctor," was tried at Nuremburg for surgical experiments on concentration camp prisoners and hanged on June 2, 1948.

35. National Archives microfilm T-77/R1429/144-145. This is a copy of Jodl's actual handwritten diary, which is difficult to read. A typed version (T-77/R1430/923)—printed by the Historical Division, Headquarters U.S. Army, Europe, Foreign Military Studies Branch 1943–45—has a question mark in parentheses after the word *Weltraumschaff* (spaceship), perhaps because it was misspelled (*Weltraumschiff* would be more correct) or perhaps due to some kind of incredulity. In the handwritten text, there is a pen scratch above the word that might be Jodl's own question mark. See also Volkhard Bode and Gerhard Kaiser, *Raketenspuren* (Bechtermünz Verlag, 1998), pp. 120–21.

36. Dornberger, p. 207.

37. Speer, *Slave State*, pp. 71 and 94–95. Speer found that "minor employees" acting as SD agents "vented their annoyance at their superiors in transparent suspicions." He was aware that "they regarded it as their bound duty to find unpleasant things to report at any price." This showed "the nature and danger of any system of informants." Even Martin Bormann lost patience with the SD, fuming that "completely irresponsible people make allegations and accusations, while the responsible people are not even consulted" (ibid., p. 106).

38. CIC Agent Report dated April 1, 1953, a summary of an interview with von Braun on March 30, 1953, stated: "Himmler, not having made any progress in making himself the master of the V-2 Project, had SUBJECT, HIS brother, and other leading members of the Project apprehended by the Gestapo in February 1944. They were held prisoner in Stettin, Germany. HE was charged with co-operation with the Western Allies with claiming that HE wanted to use the V-2 as an instrument of science rather than as a weapon. They also accused HIM and HIS brother of desiring to escape to England and to sabotage the V-2 Project at Peenemuende. They were finally released through the intervention of the Wehrmacht Military Intelligence" (Author's files via FOIA).

39. Long-lasting suspicions that the gestapo sabotaged Riedel's car have never been proven. Von Braun apparently took him off the test stands because he was an unsystematic record keeper (Reisig interview, National Air and Space Museum, June 5, 1989, p. 55). The town of Bernstadt, Germany, triggered controversy in 2008 when it named a school after him. See "The V2 kindergarten: British fury as Germans name school after maker of WWII terror rocket," *London Daily Mail*, February 5, 2008. A crater on the moon was named in his honor in 1970.

40. The Red Army finally took over the Norhausen area on July 1, 1945. The Soviets were as eager to acquire rocket experts as the other Allies, offering generous food rations and high salaries to thousands. At first they kept the work in Germany, but in October 1946 suddenly deported the men and their families en masse to Russia. As Clarence Lasby showed, the "Americans had not used force, but they *had* removed large numbers of scientific personnel; the Russians *had* used force, but they could cite legal justification" in a September 1945 Allied proclamation that directed German authorities "to provide equipment, materials, and personnel *for use in Germany or elsewhere* as the Allied representatives might direct" (*Project*

Paperclip, p. 183). In any case, for at least their first few years in America, the Germans were essentially prisoners. Regarding Grötrupp, Donald MacKenzie concluded that he and several other German guidance specialists "probably would have been insufficient to sustain a major guidance development effort" for the Soviets, who already possessed a "high level of indigenous technical expertise." The Soviet Union "had both the capacity and the desire to learn from German work, not merely to copy it." Donald MacKenzie, *Inventing Accuracy* (MIT Press, 1990), pp. 303–7. By the early 1950s the Soviets were sending their German rocket men home—Grötrupp and his family returned to Germany in 1953. McGovern, pp. 231–32.

41. Lasby, pp. 45–46.
42. Ibid., p. 79. The military exploitation program would become "Project Paperclip," which included commercial-industrial programs, in March 1946.
43. Hunt.
44. Lang, pp. 87–88.
45. Author's files via FOIA.
46. Author's files via FOIA. On the "Basic Personnel Record" (fn. 29), von Braun's annual income at Peenemünde was noted as 31,000 marks, which suggests that the U.S. Army gave him a raise of 200 marks. In 1945, the Allies introduced an exchange rate of 10 marks to the dollar. Based on the Consumer Price Index, his salary would be worth about $36,000 today. In 1945, this would have made him as economically powerful as someone earning $193,000 today. For these calculations, see www.history.ucsb.edu/faculty/marcuse/projects/currency.htm and www.measuringworth.com.
47. See Hunt, *Secret Agenda*, pp. 60–61, for a discussion of tampering with the Mittelwerk documents. Details of the journey from Germany to Texas have been filled in by voluminous worshipful writing about von Braun over the years, much of it generated by himself. I have chosen not to traffic in them here. James McGovern's 1965 *Crossbow and Overcast*, published in London, was the first objective account using documented sources, but he relied heavily on interviews with Dornberger and von Braun that parroted their earlier writing. One of the more believable anecdotes is that Hamill changed their train seats in St. Louis when he discovered they had been assigned to a car full of wounded soldiers. For an early example of popular coverage, see "We Want with the West," *Time*, December 9, 1946.

EPILOGUE

1. See obituary of Francesco "Pancho" Morales, *Washington Post*, January 5, 1997, p. B8.
2. Photo in David H. DeVorkin, *Science with a Vengeance* (Springer-Verlag, 1992), p. 46.
3. See John Gimbel, "Project Paperclip: German scientists, American policy, and the cold war," *Diplomatic History* 14 (1990), pp. 343–65.

4. They were also gathered at Fort Strong, Boston, where twenty-eight of them, including Arthur Rudolph, signed a handmade four-page illustrated Christmas card from the "Deutsche Wissenschaftler in Fort Strong" expressing "Frohe Weihnachten und ein Glückliches Neues Jahr" (Merry Christmas and Happy New Year) and "Wünschen" (Best Wishes) to their captors. I wish to thank Professor Arno J. Mayer for his generosity in lending me the card.

5. See DeVorkin, chapters 5–8 for a discussion of this work.

6. Daniel Lang, "A Romantic Urge," *The New Yorker*, April 21, 1951, p. 89.

7. *El Paso* (Texas) *Times*, December 6, 1946.

8. Lang, p. 89.

9. See Clarence G. Lasby, *Project Paperclip* (Athenum, 1971), chapter 5, for a discussion of the public reaction against exploiting the Germans.

10. Quoted in Lasby, p. 190. F.A.S. opposition to bringing German scientists to the United States had supporters in the State Department, who wanted meticulous scrutiny of émigrés (Gimbel, p. 362).

11. Lasby, p. 193.

12. Quoted in "German Scientists in El Paso Blasted," *El Paso* (Texas) *Times*, July 1, 1947.

13. "Office of the Military Government of the U.S., Revised Security Report on German (or Austrian) Scientist or Important Technician." The report did note somewhat ambiguously that von Braun "was an SS officer but no information is available to indicate that he was an ardent Nazi." By the time of this report, central records of the German government had become available, thus providing accurate dates for his Nazi Party and SS membership. Author's collection via FOIA.

14. Gimbel maintained that "the emphasis on the Russians that appeared frequently in the public forum may be interpreted as a convenient rationale to make the [Paperclip] program acceptable to the public at large" (Gimbel, pp. 364–65).

15. Gimbel called "a significant loophole indeed" the language in a revised Paperclip directive of August 1946 that nullified the disqualification of Nazi Party members or active supporters of Nazi militarism if the position or honors "awarded a specialist under the Nazi regime" were "solely on account of his scientific or technical ability." Individual cases "made a mockery of American denazification policies," Gimbel wrote (Gimbel, pp. 356–57).

16. Quotation from personal letter to the author dated March 17, 1993.

17. *The New Yorker* referred to Maria in 1951 as von Braun's "second cousin he had known all his life" (Lang, p. 77).

18. DeVorkin, p. 181.

19. Lang, p. 79. "By the way," the *New Yorker* writer was told by one of von Braun's Army handlers, "the group has been investigated thoroughly by both the Army and the F.B.I. You never know when someone's going to ask about that." He said that von Braun had joined the Nazi Party in 1940—wrong by three years—"more or less as a matter of expediency." There was of course no way for the magazine to fact-check this information.

20. In a letter dated September 11, 1952 to a member of the British Interplanetary Society, von Braun wrote: "Instead of *my* novel (which turned out to be highly controversial from the publisher's point of view) the Bechtle Verlag is going to publish a novel written by Mr. F. L. Neher, which is at least partly based on the scientific and technical material presented in my scientific booklet. Outside of the fact that I have sold the rights to use my material to Bechtle (and that I shall write a preface for Neher's book), I have no connections with this project anymore" (Von Braun to Gatland, Wernher von Braun papers, Library of Congress, box 42). The book was published in 1953 as *Menschen zwischen den Planeten*, by Franz Ludwig Neher. See George Mandler, *Interesting Times: An Encounter With the 20th Century 1924–* (Lawrence Erlbaum Associates, 2001), pp. 118–22, for background on Neher.

21. Lang, p. 92.

22. Willy Ley helped to write the *Collier's* articles. His comments are in Wernher von Braun papers, Library of Congress, box 42, first "Collier's" file.

23. I am grateful to journalist Michael Wright, who grew up in Alabama during the 1950s, for telling me about this phrase.

24. For detailed discussion of the American and Soviet satellite programs before and after *Sputnik*, see Robert A. Divine, *The Sputnik Challenge* (Oxford University Press, 1993), and Roger D. Launius et al., eds., *Reconsidering Sputnik* (Harwood Academic, 2000).

25. See Wayne Biddle, "AAAS's very own Red Scare," *Science* 260 (April 23, 1993), p. 486. The army also advised him not to mix with the Academy of American Poets. He was permitted to speak before conservative business and religious groups.

26. Jennings to Frazer, December 15, 1958, Wernher von Braun papers, Library of Congress, box 3.

SELECTED BIBLIOGRAPHY

Abraham, David. *Collapse of the Weimar Republic*. Holmes Meier, 1988.

Ahrweiler, Jacques. *Le Cas von Braun*. Editions Seghers, 1972.

Argoux, Georges, et al. *Le camp de concentration de Dora, ses Kommandos et leur cadre général*. Presse d'Aujourd'hui, 1989.

Asprey, Robert B. *The German High Command at War*. William Morrow, 1991.

Bach, Jurgen A. *Franz von Papen in der Weimarer Republik*. Droste, 1977.

Beevor, Anthony. *The Fall of Berlin 1945*. Viking Press, 2002.

Béon, Yves. *Planet Dora*. Westview Press, 1997.

Berdahl, Robert M. *The Politics of the Prussian Nobility*. Princeton University Press, 1988.

Bergaust, Erik. *Wernher von Braun*. National Space Institute, 1976.

Bessel, Richard. *Germany After the First World War*. Clarendon Press, 1995.

Bessel, Richard, and Edgar Feuchtwanger, eds. *Social Change and Political Development in Weimar Germany*. Croom Helm Ltd., 1981.

Biddle, Wayne. "AAAS's Very Own Red Scare." *Science* 260 (1993).

———. "Science, Morality and the V-2." *The New York Times* (October 2, 1992).

———. "A Great New Enterprise." *Air & Space Smithsonian* (June/July 1989).

Blackbourn, David, and Richard J. Evans, eds. *The German Bourgeoisie*. Routledge, 1993.

Bloch, Michael. *Ribbentrop*. Crown, 1992.

Bode, Volkhard, and Gerhard Kaiser. *Raketenspuren, Peenemünde 1936–1996*. Bechtermünz Verlag, 1998.

Bower, Tom. *The Paperclip Conspiracy*. Little Brown, 1987.

Bracher, Karl Dietrich. *The German Dictatorship*. Praeger, 1970.

Broszat, Martin. *Hitler and the Collapse of Weimar Germany*. Berg, 1987.

Brustein, William. *The Social Origins of the Nazi Party, 1925–1933*. Yale University Press, 1996.

Bull, Gerald V., and Charles H. Murphy. *Paris Kanonen—The Paris Guns*. Presidio Press, 1991.

Calvocoressi, Peter, Guy Wint, and John Pritchard. *Total War: The Causes and Courses of the Second World War*. Pantheon, 1989.

Cameron, Norman, and R. H. Stevens, trans. *Hitler's Table Talk 1941–1944*. Enigma Books, 2000.

Craig, Gordon A. *The Germans*. Meridian, 1983.

———. *Germany 1866–1945*. Oxford University Press, 1980.

———. *The Politics of the Prussian Army, 1640–1945*. Oxford University Press, 1964.

DeVorkin, David H. *Science with a Vengeance*. Springer-Verlag, 1992.

Dietrich, Axel. *Peenemünde through the Centuries*. Axel Dietrich Verlag, 1994.

Divine, Robert A. *The Sputnik Challenge*. Oxford University Press, 1993.

Dornberger, Walter. "The German V-2." *Technology and Culture* (Fall 1963).

———. *V-2*. Viking Press, 1954.

Dorpalen, Andreas. *Hindenburg and the Weimar Republic*. Princeton University Press, 1964.

Duffy, Christopher. *Red Storm on the Reich*. Atheneum, 1991.

Evans, Richard J. *The Coming of the Third Reich*. Penguin Press, 2004.

Feldman, Gerald D. *The Great Disorder: Politics, Economics, and Society in the German Inflation, 1914–1924*. Oxford University Press, 1997.

Fest. Joachim C. *Hitler*. Mariner Books, 2002.

———. *Speer: The Final Verdict*. Harcourt, 2001.

———. *The Face of the Third Reich*. Pantheon, 1970.

Feuchtwanger, Edgar. *Imperial Germany 1850–1918*. Routledge, 2001.

———. *From Weimar to Hitler*. St. Martin's Press, 1993.

Fritzsche, Peter. *A Nation of Flyers: German Aviation and the Popular Imagination*. Harvard University Press, 1994.

———. "Machine dreams: Airmindedness and the reinvention of Germany." *American Historical Review* 98, no. 3 (June 1993).

Gall, Lothar. "Introduction." In *Questions on German History, an Historical Exhibition in the Deutscher Dom Berlin*. German Bundestag, 1998.

Garnier, Louis, and Jean Mialet. "Preface." In *Dora, la mangeuse d'hommes: Reproductions de 35 lavis faits en 1945 par Maurice de la Pintière*. Presse d'Aujourd'hui, 1993.

Gay, Peter. *Weimar Culture*. W. W. Norton, 2001.

———. *The Cultivation of Hatred*. W. W. Norton, 1993.

Generales, Constantine D. J. "Wernher von Braun." *New York State Journal of Medicine* (November 1977).

Generales, Constantine. "Recollections of early biomedical moon-mice investigations." *Smithsonian Annals of Flight*, no. 10 (1974).

Gilens, Alvin. *Discovery and Despair: Dimensions of Dora*. Westkreuz Verlag, 1995.

Gimbel, John. *Science, Technology and Reparations: Exploitation and Plunder in Postwar Germany.* Stanford University Press, 1990.

———. "Project Paperclip: German scientists, American policy, and the cold war." *Diplomatic History* 14 (1990).

Graves, Robert. *Good-bye to All That.* Anchor Books, 1985.

Halperin, S. William. *Germany Tried Democracy: A Political History of the Reich from 1918 to 1933.* W. W. Norton, 1974.

Harrison, Mark. "A Soviet quasi-market for inventions: Jet propulsion, 1932–1946." *Research in Economic History* 23 (Elsevier, 2005).

Herf, Jeffrey. *Reactionary Modernism.* Cambridge University Press, 1990.

Hess, Torsten, Thomas A. Seidel, et al. *Vernichtung durch Fortschritt: Am Bespiel der Raketenproduktion im Konzentrationslager Mittelbau.* Westkreuz, 1995.

Hetzel, Robert L. "German monetary history in the first half of the twentieth century." *Economic Quarterly* (January 1, 2002).

Höhne, Heinz. *The Order of the Death's Head.* Penguin Books, 2000.

Holborn, Hajo. *A History of Modern Germany, 1840–1945.* Princeton University Press, 1982.

Hölsken, Heinz Dieter. *Die V-Waffen.* Deutsche Verlags-Anstalt, 1984.

Hunt, Linda. *Secret Agenda.* St. Martin's Press, 1991.

———. "U.S. Coverup of Nazi Scientists." *Bulletin of the Atomic Scientists* (April 1985).

Irving, David. *The Mare's Nest.* Little Brown, 1965.

James, Harold. *The German Slump: Politics and Economics 1924–1936.* Clarendon Press, 1986.

Johnson, David. *V-1, V-2: Hitler's Vengeance on London.* Stein & Day, 1981.

Johnson, Stephen B. *The Secret of Apollo: Systems Management in American and European Space Programs.* Johns Hopkins University Press, 2002.

Jones, Larry Eugene. " 'The greatest stupidity of my life': Alfred Hugenberg and the formation of the Hitler cabinet, January 1933." *Journal of Contemporary History* 27 (1992).

Kaes, Anton, et al. eds. *The Weimar Republic Sourcebook.* University of California Press, 1994.

Kershaw, Ian. *Hitler 1936–1945: Nemesis.* W. W. Norton, 2001.

———. *Hitler 1889–1936: Hubris.* W. W. Norton, 2000.

Kracauer, Siegfried. *From Caligari to Hitler: A Psychological History of the German Film.* Princeton University Press, 2004.

Lang, Daniel. "A Romantic Urge." *The New Yorker* (April 21, 1951).

Lasby, Clarence G. *Project Paperclip.* Atheneum, 1971.

Launius, Roger D., ed. *Innovation and the Development of Flight.* Texas A&M University Press, 1999.

Launius, Roger D., et al., eds. *Reconsidering Sputnik.* Harwood Academic, 2000.

Ley, Willy. *Events in Space.* David McKay Company, 1969.

———. "Count von Braun." *Journal of the British Interplanetary Society* 6 (June 1948).

Lindemann, Albert S. *Esau's Tears*. Cambridge University Press, 1997.

Ludwig, Karl-Heinz. *Technik und Ingenieure im Dritten Reich*. Droste, 1974.

Lukacs, John. *The Hitler of History*. Alfred A. Knopf, 1997.

Lutz, Ralph Haswell, ed. *Fall of the German Empire, 1914–18*. Stanford University Press, 1932.

MacKenzie, Donald. *Inventing Accuracy*. MIT Press, 1990.

———. "The Soviet Union and strategic missile guidance." *International Security* 13, no. 2 (Fall 1988).

Mader, Julius. *Geheimnis von Huntsville: Die wahre Karriere des Raketenbarons Wernher von Braun*. Deutscher Militärverlag, 1963.

Mailer, Norman. *Of a Fire on the Moon*. Little Brown, 1970.

Marshall, S.L.A. *World War I*. Houghton Mifflin, 1987.

Mayer, Arno J. *The Persistence of the Old Regime*. Pantheon, 1981.

McGovern, James. *Crossbow and Overcast*. Hutchinson, 1965.

Mialet, Jean. "Dora, le camp oublié." *Bulletin de l'Amicale des Prisonniers Politiques de Dora* (Spring 1995).

Middlebrook, Martin. *The Peenemünde Raid*. Penguin Books, 1988.

Mosse, George L. *The Crisis of German Ideology: Intellectual Origins of the Third Reich*. Howard Fertig, 1999.

———. *Germans and Jews: The Right, the Left, and the Search for a Third Force in Pre-Nazi Germany*. Wayne State University Press, 1987.

Muncy, Lysbeth Walker. *The Junker in the Prussian Administration under William II, 1888–1914*. Brown University Press, 1944.

Murray, Bruce. *Film and the German Left in the Weimar Republic: From Caligari to Kuhle*. University of Texas, 1990.

Neufeld, Michael J. *Von Braun: Dreamer of Space, Engineer of War*. Alfred A. Knopf, 2007.

———. "Wernher von Braun, the SS, and concentration camp labor: Questions of moral, political, and criminal responsibility." *German Studies Review* 25, no. 1 (2002).

———. *The Rocket and the Reich*. Free Press, 1995.

———. "Hitler, the V-2, and the battle for priority, 1939–1943." *The Journal of Military History* no. 3 (July 1993).

———. "Weimar culture and futuristic technology: The rocketry and spaceflight fad in Germany, 1923–1933." *Technology and Culture* 31 (1990).

Ordway, Frederick I. III, and Mitchell R. Sharpe. *The Rocket Team*. Thomas Crowell, 1979.

Overy. Richard, *Interrogations: The Nazi Elite in Allied Hands, 1945*. Penguin Books, 2001.

———. *Goering*. Phoenix Press, 2000.

———. *Why the Allies Won*. W. W. Norton, 1995.

———. *War and Economy in the Third Reich*. Clarendon Press, 1994.

Padfield, Peter. *Himmler*. Henry Holt, 1990.

Peukert, Detlev J. K. *The Weimar Republic*. Hill and Wang, 1992.

Powers, Thomas. *Heisenberg's War*. Alfred A. Knopf, 1993.

Proctor, Robert N. *Value-Free Science?*. Harvard University Press, 1991.

Pynchon, Thomas. *Gravity's Rainbow*. Bantam Books, 1974.

Renneberg, Monika, and Mark Walker, eds. *Science, Technology and National Socialism*. Cambridge University Press, 1994.

Richie, Alexandra. *Faust's Metropolis*. Carroll & Graf, 1998.

Roos, Hans. *A History of Modern Poland*. Alfred A. Knopf, 1966.

Ruland, Bernd. *Wernher von Braun: Mein Leben für die Raumfahrt*. Burda Offenburg, 1969.

Schacht, Hjalmar. *My First Seventy-six Years*. Allan Wingate, 1955.

Schmidt, Matthias. *Albert Speer, the End of a Myth*. St. Martin's Press, 1984.

Sereny, Gitta. *Albert Speer: His Battle with Truth*. Alfred A. Knopf, 1995.

Shriver, Donald W. *Honest Patriots*. Oxford University Press, 2005.

Simpson, Christopher. *Blowback*. Weidenfeld & Nicolson, 1988.

Speer, Albert. *The Slave State*. Weidenfeld & Nicolson, 1981.

———. *Inside the Third Reich*. Macmillan Company, 1970.

Spotts, Frederic. *Hitler and the Power of Aesthetics*. Overlook Press, 2002.

Stackelberg, Roderick. *Hitler's Germany: Origins, Interpretations, Legacies*. Routledge, 1999.

Stuhlinger, Ernst, and Frederick I. Ordway. *Wernher von Braun, Crusader for Space*. Krieger Publishing Company, 1994.

The Treaty of Versailles and After. Washington, DC: U.S. Government Printing Office, 1947.

Turner, Henry Ashby. *German Big Business and the Rise of Hitler*. Oxford University Press, 1985.

von Braun, Magnus. *Von Ostpreussen bis Texas*. Stollhamm, 1955.

von Braun, Wernher. *Recollections of Early Childhood/Early Experiences in Rocketry, As Told by Wernher von Braun, 1963*, Marshall Space Flight Center.

———. "Space man—the story of my life." *American Weekly* (1958).

———. "Reminiscences of German rocketry." *Journal of the British Interplanetary Society* 15, no. 3 (May–June 1956).

von Hochberg, Stephanie, and Holger Steinte. " 'Von der Hölle zu den Sternen': Wernher von Braun, die Entwicklung der Rakete und das 'Dritte Reich.' " *Ich diente nur der Technik*. Museum für Verkehr und Technik Berlin, 1995.

Wagner, Jens-Christian. *Produktion des Todes Das KZ Mittelbau-Dora*. Wallstein Verlag, 2001.

Ward, Bob. *Dr. Space: The Life of Wernher von Braun*. Naval Institute Press, 2005.

Wegener, Peter P. *The Peenemünde Wind Tunnels, A Memoir*. Yale University Press, 1996.

Weitz, Eric D. *Weimar Germany: Promise and Tragedy*. Princeton University Press, 2007.

Weyer, Johannes. *Wernher von Braun*. Rowohlt, 1999.

Wheeler-Bennett, John. *The Nemesis of Power*. Macmillan, 1954.

Willett, John. *Art and Politics in the Weimar Period*. Pantheon, 1978.

Williamson, John G. *Karl Helfferich*. Princeton University Press, 1971.

Winter, Frank H. *Rockets into Space*. Harvard University Press, 1990.

Zembsch-Schreve, Guido. *Pierre Lalande: Special Agent*. Leo Cooper, 1996.

PHOTOGRAPH CREDITS

Page 32: NASA Marshall Space Flight Center Collection

1. NASA Marshall Space Flight Center Collection
2. NASA Marshall Space Flight Center Collection
3. Ullstein bild / The Granger Collection, New York
4. Courtesy of Wayne Biddle
5. KZ-Gedenkstaette Mittelbau-Dora
6. National Archives and Records Administration
7. U.S. Army / Time & Life Pictures / Getty Images
8. National Archives and Records Administration
9. National Archives and Records Administration
10. National Archives and Records Administration
11. AP / Wide World Photos
12. National Archives and Records Administration
13. NASA Marshall Space Flight Center Collection
14. NASA Marshall Space Flight Center Collection
15. NASA Marshall Space Flight Center Collection
16. NASA Marshall Space Flight Center Collection
17. NASA Marshall Space Flight Center Collection
18. NASA Marshall Space Flight Center Collection

INDEX

Page numbers beginning with 153 refer to notes.
Page numbers in *italics* refer to illustrations.

Aachen, Technical University of, 91, 181
Advance into Space (Valier), 40–41
Ahlbeck, 84, 178
Alexander I, Czar of Russia, 9
All Quiet on the Western Front (Remarque),
  55
anti-Semitism, 8, 43, 156, 159, 165
  Austrian, 106
  Nazi, 5, 6, 7, 41, 55, 58, 59, 64, 73–74, 89,
    154, 166
Apollo Project, xii, 127–28
Ariernachweis ("Aryan certificate"), 94
atomic bomb, x, 38, 130–31, 135, 144, 145,
  164, 192
Atwood, J. Leland, 147–48
Auf zwei Planeten (Lasswitz), 62, 170
Auschwitz concentration camp, 70, 121,
  187
Austria, 9, 101, 105–6
Austria-Hungary, 11, 17–18, 158
Austro-Prussian War (1866), 20
aviation pioneers, 34, 38, 44–45, 54, 114,
  164, 168, 181

Bauhaus style, 87, 179
Bäumker, Adolph, 82

Bavaria, 30, 39–40, 41, 59
Becker, Karl, 66, 67, 68, 79, 82, 83, 92, 105,
  171
  Nebel and, 44, 50, 51, 53
  suicide of, 74, 113, 174
  university-military partnership
    established by, 69, 73, 174
Benjamin, Walter, xv
Bergen-Belsen concentration camp, 122,
  132
Berger, Gottlob, 112, 184
Berlin, 2, 4, 12, 22, 27, 28, 31, 40, 45, 117, 129
  AVUS racetrack, 41, 48, 165
  Deutsche Bank, 21
  trade minister's palace, 13–14, 157
  violence in, 5, 59, 67, 73–74
  worker uprisings in, 29
Berlin, Technical University of, 37, 44,
  51–52, 69, 168
  Nazi activity in, 52, 60, 168
Berlin, University of, 73–74, 176
Berliner Tageblatt, 26
Bernhard, Georg, 26
Bethe, Hans, 146–47
Bethmann-Hollweg, Theobald von, 22,
  23, 24, 25, 156–57, 160, 161

Bismarck, Otto von, 20, 156, 164
black-market activity, 22, 23, 159
Blomberg, Werner von, 101–2, 182
Blut und Boden (blood and soil) ideology,
    65
Boer War, 14, 155
Bormann, Martin, 76, 194
Borsig, Ernst von, 52
Borsig Locomotive Factory, 52, 68
Boykow, Johannes Maria, 91, 99, 100,
    180–81
Brauchitsch, Walther von, 105, 106, 107,
    184
Braun, Wernher von, *see* von Braun,
    Wernher Magnus Maximilian
    Freiherr
Brill, Dorothee, 117–18, 186
British Interplanetary Society, 148, 197
Bruning, Heinrich, 49, 56–57, 63, 64, 168
Buchenwald concentration camp, 70, 115,
    121, 122, 124, 125, 126, 186–87, 189
Busch, Wilhelm, 79

Casablanca Conference (1943), 111, 184
Catholic Center Party, 36, 49, 155, 168, 171
Cazabonne, Robert, 125–26
Chamberlain, Neville, 106–7
Chemische-Technische Reichsanstaldt,
    52–54
Churchill, Winston, 129, 132, 173, 184
Collier's, 149–50, 197
Communists, 38, 64, 138, 141, 155, 168, 172
    alleged Jewish-Bolshevist conspiracy
        and, 29–30, 89, 109–10
    Nazi clashes with, 5, 41, 59, 60
    Rotfront, 29, 59
concentration camps, 6, 70, 102, 107, 155,
    169, 186, 193
    forced labor factories of, 111, 114–17,
        120–21, 185
    Mittelbau-Dora prisoners evacuated to,
        132–33
Conrad III, King of Germany, 8
Coolidge, Calvin, 154
Cornell University, 146–47

Crow, Sir Alwyn Douglas, 142
Czechoslovakia, 101, 106–7

Daladier, Édouard, 106
Darré, Walter, 65, 77
Delbruck, Clemens von, 13–15, 22–24,
    156–57
Depression era, 2, 5–6, 52, 56–58, 62, 65,
    89, 170
Deutsche Allgemeine Zeitung, 40
Deutsche Bank, 21, 24, 160
Deutscher Luftsport-Verband (German
    Sport Aviation Club), 178–79
Deutschnationale Volkspartei (DNVP)
    (German National People's Party),
    43, 160, 161, 171
Dietrich, Hermann, 31
Dingell, John D., 147
Disney, Walt, xi, 128, 150
Dora, *see* Mittelbau-Dora slave labor
    camp
Dornberger, Walter, 66–67, 68, 81, 84, 90,
    93, 97, 100, 116, 121, 133, 135, 136,
    139, 142, 162–63, 166–67, 170,
    171–72, 173, 174, 175, 176–77, 178,
    183, 185, 186, 187, 192
    "all under one roof" organizational
        scheme of, 80, 110
    bird hunting by, 94, 98
    fund-raising efforts of, 73, 76, 101,
        104–5, 112–13, 184
    Himmler's visit described by, 118–19
    Hitler meeting requested by, 111–13,
        117, 184
    at Hitler's Kummersdorf visit, 107–8
    at Kummersdorf, 77–78, 79–80, 82, 83,
        176–77
    military rocket program headed by, 72,
        82, 83, 88, 95
    multimedia lecture shows of, 80, 83, 117,
        119–20, 177
    Paris Gun and, 76, 174
    supersonic wind tunnel sought by,
        91–92
    surrender to U.S. Army by, 128–37, 190

test-stand explosion recalled by, 77–78,
        175
    U.S. immigration of, 148
    WVB as described by, 87–88, 91, 104
    on WVB's gestapo arrest, 139, 140, 141
    on WVB's SS membership, 74–75
Douglas, Lord Alfred, 4, 154
Dreadnought, HMS, 11
Durant, Frederick C., III, 180

East Prussia, 7, 8–10, 17, 31, 117, 157, 170,
        174
    in World War I, 18–21, 158
Ebert, Friedrich, 28, 39, 57, 169
Edward VII, King of England, 12
Ehrhardt brigade, 39
Einstein, Albert, ix, x, 13, 78, 147
    WVB's letter from, 60
Eisenhower, Dwight D., x, 129, 143, 144,
        151
Enabling Act, 6, 155
Engel, Rolf, 172
Enlightenment, 37
Erzberger, Matthias, 169
Ettersburg Castle, 48, 167

Fatherland Party, 30, 161
Federation of American Scientists, 147
Forge, John, 153
FP1 antwortet nicht ("FP1 Is Not Respond-
        ing"), 98
France, 9, 10, 11, 34, 40, 80, 166, 174, 177,
        178
    slave labor prisoners from, xiii, 71,
        115–17, 121, 124–26, 189
    in World War I, 18, 20, 28, 158
    in World War II, 82, 101, 106–7, 109, 113
Francis Ferdinand, Archduke of Austria,
        17–18
Franco, Francisco, 89
François, Hermann von, 19
Frau im Mond (*Woman in the Moon*), 42,
        46–47, 48, 62, 91, 149, 166
    Oberth and, 43, 44, 51, 52–54, 55, 79
    technical effects of, 43, 119

Freikorps, 29, 30, 35, 73, 161, 162, 163
Frick, Wilhelm, 76
Friedrich-Wilhelm University, 69, 78, 172
Fritsch, Werner Freiherr von, 83, 105, 177
    in homosexuality scandal, 101–2, 182

Gebhardt, Karl, 140, 193
Generales, Constantine "Ntino," 60–61,
        170
German Democrats, 36
German National People's Party
        (*Deutschnationale Volkspartei*)
        (DNVP), 43, 160, 161, 171
Germany, Imperial, 11, 12, 17–28
    abject monarchism in, 3, 14
    food shortages in, 21–22, 23, 25, 28, 159,
        160–61
    Interior Ministry of, 22, 23–27, 159,
        160–61
    science in, 12–13
    Trade Ministry of, 13–15, 22
    wartime economy of, 21–22, 23–24, 28,
        29, 159, 160
    see also World War I
Germany, Nazi, x, xi, 4–7, 38, 71, 75, 76,
        77, 79, 83, 85, 97, 98, 100–101, 106,
        109–10, 114, 115, 118, 121, 136, 162,
        176, 187
    Air Ministry of, 82, 83
    Navy of, 81–91, 177, 180
    nuclear energy program of, 176, 192
    population of, 89
    racist policies of, 94–95, 117–18, 181
    Saar region plebiscite and, 80
    Soviet Union invaded by, 109, 132
    see also Hitler, Adolf; Nazi Party;
        Nazis
Germany, Weimar Republic of, 4–7,
        28–31, 33–45, 46–55, 56–69, 76–77,
        89, 105
    antidemocracy sentiment in, 35–36
    anti-Republic sentiment in, 30, 35,
        39–40, 42–43, 45, 51, 56–58, 63,
        162, 165
    Constitution of, 155, 157, 161, 168

Germany, Weimar Republic (*continued*)
  economic problems of, 5, 28–29, 39, 40,
      42, 43–44, 53, 54, 56–59, 62–63, 77,
      168, 169
  estate owners in, 60, 64, 65, 170
  hyperinflation in, 39, 46, 163
  Kapp putsch in, 30–31, 36, 43, 50,
      161–62
  navy of, 180
  1932 elections of, 62–64, 98, 171
  as parliamentary democracy, 34, 56, 63
  right-wing groups in, 30–31, 39–40,
      49–51, 52, 55, 59, 157, 162, 182
  unemployment in, 44, 45, 49, 55, 56, 57,
      59, 165, 166, 168
  Versailles Treaty opposed in, 34–36, 50,
      67–68
  war reparations owed by, 34, 50–51, 63,
      166, 168
  worker uprisings in, 29–30, 31
  see also Versailles, Treaty of
gestapo, 78–79, 82–83, 124, 125, 138–41,
      193, 194
Goddard, Robert, 36, 62, 171
Goebbels, Joseph, 44, 57, 63, 73–74, 107,
      154, 155, 165, 169
  Propaganda Ministry of, 78, 161, 173
Goethe, Johann Wolfgang von, 11, 156,
      167
Göring, Hermann, 44, 57, 76, 81, 86, 89,
      94, 101, 102, 140, 157, 177, 182
Gostkowsky, Eleonore von, 9
Graves, Robert, 162, 163
Great Britain, 12, 14, 15, 28, 80, 101, 145,
      158, 178, 180, 181, 183
  banking system of, 11–12
  Boer War of, 14, 155
  Dornberger interned in, 142
  German undercover agents in, 143
  Munich Conference and, 106–7
  Royal Air Force of, 111, 115, 116, 117, 185
  V-1 attacks on, 173
  V-2 attacks on, 130, 142–43, 190
  WVB interrogated in, 142–43

Greifswalder Oie test launch site, 95, 96,
      97–100, 107, 119
  Deutschland A-3 launch event at,
      98–100, 102, 103
Guernica, bombing of, 93, 94, 177
Gumbinnen, 29, 30–31
Gürtner, Franz, 64, 77, 102, 175

Haber, Fritz, 21, 85, 159, 178
Haber-Bosch process, 84–85
Hamill, James P., 144, 190–91, 195
Harris, Gordon L., 182–83
Heidelberg, University of, 39, 165
Heimat movement, 8
Heinkel aircraft, 94, 185
Heisenberg, Werner, 78, 175, 192
Helfferich, Karl, 24–25, 160
Herf, Jeffrey, 37
Hermann, Rudolf, 91, 92, 93, 100, 103
Hess, Rudolph, 76
Heylandt, Paul, 78, 165, 167, 171
Himmler, Heinrich, 6, 73, 74–76, 102, 109,
      117–19, 173, 174, 182, 192
  appearance of, 183
  armaments makers arrested by, 138–41,
      193, 194
  Austrian anti-Semitic terrorism
      organized by, 106
  marriage applications scrutinized by,
      117–18, 186
  Peenemünde and, 111–12, 117, 118–19,
      126, 184, 186
  slave labor camps established by, 114–15,
      120–21, 187
  WVB's conversation with, 118–19
Hindenberg und Beneckendorff, Paul
      Ludwig Hans Anton von, 4–5,
      20–21, 24–26, 29, 50, 57, 63, 64, 65,
      77, 158–59, 161, 168, 171
Hitler, Adolf, x, 4–7, 43, 44, 50, 51, 52, 55,
      57, 71, 82–83, 86, 94, 97, 133, 155,
      161, 163, 166, 169, 174, 183, 185, 190
  A-4 mass production ordered by, 117,
      119–20

annihilating effect desired by, 119–20, 149

architecture favored by, 87, 179

ascension to chancellorship of, 4–5, 34, 49, 62–64, 77

Austria annexed by, 105–6

"Axis" alliance formed by, 89–90

"beer hall *Putsch*" of, 39, 40, 64

Blomberg-Fritsch imbroglio and, 101–2

charismatic effect of, 180

coup d'etat assassination attempt on, 107, 111

Czechoslovakia invasion ordered by, 106–7

death of, 130

demagogy of, 58–59

Dornberger's requested meeting with, 111–13, 117, 184

"Germania" vision of, 71, 93

Kummersdorf visited by, 76, 107–8

Lebensraum vision of, 101, 109

Luftwaffe and, 80, 81

megalomania of, 109

in 1932 elections, 63, 64, 98

peace speeches of, 80

physical health of, 119

Poland invasion ordered by, 109, 113

popular appeal of, 58

potential opponents neutralized by, 6–7, 116

Rhineland invasion ordered by, 82–83, 177

rocket program downgraded by, 112–13, 114

SA purged by, 79

Spanish Civil War and, 89

Sudetenland granted to, 106

Valier and, 41

Versailles Treaty abrogated by, 76, 80

Wehrmacht controlled by, 101, 102, 105–6

World War II launched by, 71, 89–90, 100–101, 105–8, 109–10

WVB encounter with, 108, 109

WVB gestapo arrest and, 140, 193

WVB given "Professor" rank by, 119, 133

WVB's multimedia presentations to, 117, 119–20

WVB's stated impressions of, 69, 175

"Hitler's Olympics," 1936, 89

Hobsbawm, Eric, 56

Hohberg und Buchwald, Anton Freiherr von, 174

homosexuality scandal, 101–2, 182

Horstig, Ernst von, 66, 68–69

"Horst Wessel" anthem, 7, 34, 89, 155

Hugenberg, Alfred, 42–43, 47, 51, 63, 77, 166

In Stahlgewittern (Jünger), 37

"intellectual reparations," 89, 148, 180

Inter-Allied Commissions of Control, 35

intercontinental ballistic missiles (ICBMs), 113, 134, 151, 191

intermediate range ballistic missiles (IRCMs), 151

International Geophysical Year 1957–58, 151

Italy, 11, 89–90, 121–22, 178

Japan, 14, 89–90, 137, 145, 190

Jessel, Walter, 136, 138, 193

Jews, 35, 39, 57, 58, 70, 73–74, 89, 101, 156, 160, 161, 173, 178, 181, 186–87, 189

    alleged Bolshevist conspiracy of, 29–30, 89, 109–10

    English, 11

    as *kosmopolitisch,* 8

    marriage applications and, 117–18

    Polish, 22, 158, 159

    in Steglitz, 118, 186

    at trade ministry official dinners, 13–14

    see also anti-Semitism

Jodl, Alfred, 120, 139, 140, 141, 194

Jolson, Al, 59

Jouanin, Georges, 126, 189

Journal of the British Interplanetary Society,
    48
Jünger, Ernst, 37, 114, 158
Junkers, Prussian, 8–16, 43, 47, 69, 73, 148,
    160, 162, 164, 170
  class barriers maintained by, 10, 11
  Prussian officer corps filled by, 9–10, 20,
    35, 105, 158
  social milieu of, 9–10, 12, 13–15, 54
  traditional military virtues of, 9–10,
    11–12, 18, 22, 30, 31, 46, 50, 157,
    158
  see also von Braun, Magnus Alexander
    Maximilian Freiherr
Jupiter-C rocket, 151

Kafka, Franz, 118
Kalkreuth-Nieder-Siegersdorf, Count
    Eberhard, 65
Kammler, Hans, 120–21, 122, 131, 132,
    189, 191
Kant, Immanuel, 9, 10–11, 156
Kapp, Wolfgang, 30–31, 39, 42, 160, 161
Kapp putsch, 30–31, 36, 43, 50, 161–62
Kegeldüse (cone jet), 53–54, 55, 62
Keitel, Wilhelm, 120, 141, 176
Kennedy, John F., 127, 152, 180
Kershaw, Ian, 136, 180
Kesselring, Alfred, 83, 178
Keynesian economics, 57
Königsberg (Kaliningrad), 8, 10, 11, 155,
    157
Kreiselgeräte gyroscope equipment
    company, 91, 99, 103, 180
Kristallnacht pogrom, 107
Krupp, 42, 64
Kubrick, Stanley, 189
Kummersdorf proving grounds, rocket
    program of, 66–67, 76–83, 103,
    172
  A-1 rocket in, 78, 79, 92–93
  A-2 "Max" and "Moritz" rockets in,
    79–80, 83, 90, 92–93, 95, 98, 133,
    176
  A-3 rocket in, 90–91, 93

  Borkum island launch site in, 79–80, 95,
    97, 176–77
  Hitler's visits to, 76, 107–8
  Kreiselgeräte gyroscopes used by, 91, 99,
    180
  test-stand explosion at, 77–78, 175
  WVB in, 69, 71, 77–80, 81–83, 90–95,
    176

Lang, Fritz, 42–43, 44, 48, 119, 166
Lehrer, Tom, 189
Ley, Willy, 154, 166, 167, 170, 173, 181, 197
  WVB's introduction to, 46–47
Lietz, Hermann, 48, 58, 167
Lietz country boarding schools, 48–49, 51,
    58, 77, 167
Lindbergh, Charles, 41, 171
Lithuania, 8, 27, 155
"lost generation," 33, 162
Lublin extermination camp, 122
Ludendorff, Erich, 24–27, 28, 42, 158–59,
    161
Luftwaffe, 81–83, 86–87, 90, 91, 94, 102,
    119, 149, 178, 179, 184
  Condor Legion of, 93, 177
  V-1 missile of, 173, 185–86
  WVB's admiration of, 81, 82, 177
Luxemburg, Rosa, 161

Mailer, Norman, 189
Manhattan Project, 85, 97, 146, 178, 191,
    192
Mann, Thomas, 37, 114, 164
Mein Kampf (Hitler), 58
Metropolis, 42, 43
Michaelis, Georg, 25–27, 160–61
Mittelbau-Dora slave labor camp, xii–xiii,
    70–71, 72, 84, 114–26, 144, 191, 195
  executions at, 122, 124, 125, 132
  Kammler and, 120–21, 122, 189
  living conditions at, 115, 116, 121, 122
  Magnus von Braun at, 128–29, 189–90
  as Mittelwerk, 122–26, 137–38, 187
  mortality at, 115, 121, 122
  prisoner evacuation of, 130, 132–33

prisoner sabotage at, 124–25, 132
prisoner shipments to, 74, 115–16,
    121–22, 186–87
SS control of, 111, 114–16, 120–21, 122,
    124–25, 132–33
survivors' recollections of, 121, 124–26,
    185, 189, 190
in war crimes trials, 123–24, 148, 188
WVB's contacts with, 72, 115, 116–17,
    121–26, 131, 187, 188, 189, 190
Modernism, 13, 87, 157, 179
moral responsibility, scientific, ix–x, 1–2,
    85, 97, 128, 156, 173, 178
as antiprogress, 61
Morand, Guy, 124–25
Munich Conference (1938), 106–7
Mussolini, Benito, 106

Napoleon I, Emperor of the French, 9, 10,
    97, 157
nationalism, German, 8, 34, 37, 41, 42, 44,
    50–51, 58, 67–68, 77, 106
Nationalsozialistische Deutschen Studenten-
    bund (National Socialist German
    Students' League), 5, 58, 60
Nazi Party, 44, 46, 51, 52, 55, 62, 71, 72, 77,
    102, 147, 155, 168, 176, 177
banning of, 39
in Bavaria, 39–40, 41
factors in popular appeal of, 57–59
German youth in, 33, 50, 57, 58, 60,
    73–74, 169, 173–74
"Horst Wessel" anthem of, 7, 34, 89, 155
swastika emblem of, 6, 85, 87, 96, 179
WVB's membership in, 73–74, 85, 97,
    151, 181–82
Nazis:
agricultural programs of, 65, 170
anti-Semitism of, 5, 6, 7, 41, 55, 58, 59,
    64, 73–74, 89, 154, 166
anti-Young Plan campaign of, 51
banned paramilitary groups of, 63, 66
Blutkrieg (Blood War) of, 5–6, 73
coming to power of, 3–7, 159, 178
Communist clashes with, 5, 41, 59, 60

derivation of term, 154
European subjugation as goal of, 116,
    118
foreign perception of, 5
führer personality cult of, 58, 75
Kristallnacht pogrom of, 107, 181
postwar remorse lacked by, 136–37,
    192–93
propaganda of, 63, 71, 73, 89, 149, 155,
    169, 185
in Technical University of Berlin, 52, 60,
    168
torchlight parade of, 5, 154
Nebel, Rudolf, 44–45, 46, 62, 66, 69, 79,
    113, 135
dubious character of, 44, 50, 53, 92
Nebel, Rudolf, rocketry club of, 45, 46–48,
    50–55, 76
Frau im Mond rocket tested by, 52–54,
    55
Kegeldüse tested by, 53–54, 55, 62
military funding of, 50–51, 52–53,
    54–55, 62, 65–68, 171
Mirak I project of, 61–62, 66, 170
Mirak II project of, 66–67
Peenemünde researchers from, 92
Raketenflugplatz site of, 54–55, 59–60,
    61–62, 63, 64, 66–68, 72, 73, 77,
    133, 172, 175, 181
WVB in, 45, 46–48, 52–55, 59–60, 66,
    67–69, 73, 78, 133, 167, 175
Neher, Franz Ludwig, 197
New Yorker, 8, 130–31, 132, 136–37,
    142–43, 149, 191, 193
Nichomachean Ethics (Aristotle), ix
Nietzsche, Friedrich, 24, 156
Nuremberg War Crime Trials, 38, 71, 111,
    123, 142, 146, 156, 169, 175, 178,
    182, 184, 187, 188, 189, 193

Oberland, 35
Oberth, Hermann, 39–41, 43–44, 45, 49,
    51, 135, 165, 166, 167, 192
Frau im Mond rocket designed by, 43,
    44, 51, 52–54, 55, 79

Oberth, Hermann (*continued*)
   UFA funding of, 46, 55
   WVB's introduction to, 46–48
Olympic Games of 1936, 89, 131
On the Concept of History (Benjamin), xv
Opel, Fritz von, 41–42, 48, 51
Operation Barbarossa, 109, 132

Papen, Franz von, 64–65, 76–77, 171, 175
paramilitary organizations, 50, 97
   Freikorps, 29, 30, 35, 73, 161, 162, 163
   Nazi, banning of, 63, 66
"Paris Gun" (Kaiser Wilhelm Geschütz),
   28, 76, 174
Patterson, Robert, 137
Peenemünde, 74–76, 81–83, 84–95,
   96–108, 109–26, 133, 149, 178
   Allied bombing of, 111, 115, 116, 117,
      132, 137, 178, 185
   Allied forces at, 128–33, 142, 194
   architecture of, 85, 87, 112, 179
   Aryan descent required for service at,
      94
   bird hunting at, 94, 98
   funding of, 81–82, 83, 86, 87, 89, 90,
      91–92, 94, 95, 98, 101, 104–5, 177,
      185
   misleading name given to, 143
   procurement crisis at, 74, 105, 110
   reliably destructive weapons demanded
      of, 90
   research personnel of, 88, 92–93, 98,
      103–4, 109–10, 128–37, 143, 185
   secrecy of, 85, 87, 95, 110, 120
   space travel as ultimate goal of, 85–86,
      137, 139–40, 179
   supersonic wind tunnel at, 91–92, 128,
      181
   technical archives of, 131, 144, 190, 195
   WVB's charismatic effect at, 87–89,
      179–80
   WVB's critics at, 103–4, 182–83
   WVB's official position at, 72, 88, 90,
      99–100, 103
   WVB's residence at, 93–94

   see also Mittelbau-Dora slave labor camp
Peenemünde, experimental rocket
      program of:
   A-3 in, 97–100, 102, 103
   A-4 in, 100, 103, 108, 117, 118–20, 123,
      133, 134, 139–40, 150–51, 183,
      185–86, 191, 192; *see also* V-2
      rockets
   A-5 in, 100, 102, 107
   A-9 winged rocket in, 133–34
   A-10 intercontinental rocket in, 134, 191
   component drawings in, 110, 183
   engines of, 92–93, 103, 105, 107
   guidance and control work in, 103–4,
      107
   launch sequence in, 95
   specifications for, 83, 85
   testing in, 87, 117, 118–19, 134, 185–86;
      see also Greifswalder Oie test
      launch site
Peenemünde-East, 87, 100, 116, 187
Peenemünde-West, 87, 119
Peukert, Detlev, 175
Planck, Max, 12–13
Poland, 8, 16, 18, 20, 22, 34, 122, 155, 157,
      169, 178, 187
   invasion of, 109, 113, 174, 182
   Jews of, 22, 158, 159
   Mittelbau-Dora prisoners from, 74
Polanyi, Michael, ix
Preussisch-Eylau, battle of, 9
Princip, Gavrilo, 18
Prittwitz und Gaffron, Max von, 19–20,
      158
propaganda, 24, 29–30
   Nazi, 63, 71, 73, 89, 149, 155, 169, 185
   World War I, 20, 42, 159, 174
Prussia, 2, 8–16, 17–22, 34, 63, 64, 71, 73,
      97, 155, 156, 170, 171
   Interior ministry of, 11, 29
   Trade Ministry of, 13
   see also Junkers, Prussian
Prussia, Posen province of, 2, 20
   Landrat of, 15–16, 18–19, 21–22
   Polish landowners in, 16, 18, 157

quantum theory, 13
Quistorp, Emmy Melitta Cecile von, *see*
    von Braun, Emmy Melitta Cecile
    von Quistorp

racism, 5, 6, 8, 43, 58, 106, 138
    "Aryan," 72–73, 86–87, 94, 154
    as government policy, 94–95, 117–18,
        181
    see also anti-Semitism
Raiffeisen credit union, 31, 36, 163
Rakete, Die (The Rocket), 49
Raketenflug (Rocket Flight) (Nebel), 62
Rathenau, Walter, 39, 59, 169
reactionary modernism, 37, 62, 164
Redstone rocket, 150–51, 152
Rees, Eberhard, 143, 191
Reichslandbund (Rural League), 65
Reichstag, 6–7, 25–26, 28, 36, 49–50, 55,
        59, 63, 64–65, 93, 102, 155
    burning of, 6
    1926 censorship law passed by, 42–43
Reichswehr, 20, 30, 31, 35, 44, 68–69, 73,
        76, 97, 162, 163, 174, 178
Reitsch, Hanna, 169
Remarque, Erich, 55
"Republic of Science, The" (Polanyi), ix
"Responsibility and the Scientist" (Forge),
        153
Richthofen, Wolfram Freiherr von, 81–82,
        83, 93, 94, 177
Rickhey, Georg, 123, 188, 189
Riedel, Klaus, 53–54, 61–62, 67, 68, 79, 141,
        143, 167, 194
Riedel, Walter, 78, 83, 90, 93, 165, 166, 167,
        176, 191
rocket cars, 41–42, 48, 78, 165–66, 171
Rocket into Interplanetary Space, The
        (Oberth), 39–40, 49
rocketry, early history of, 33–45, 46–49,
        167
    aviation pioneers vs., 34, 38, 44–45, 164
    Frau im Mond technical effects in, 43,
        119
    right-wing associations of, 39–40, 45

rocket cars in, 41–42, 48, 78, 165–66, 171
science fiction in, 38, 39, 40–41, 42, 44,
        48, 62
space travel enthusiasts in, 38, 39–41, 42,
        43–44, 45, 47, 66, 133, 166, 171, 172
technics ideology and, 36–37, 39, 43
see also Nebel, Rudolf, rocketry club of
rocketry, German military development
        of, 34, 35, 38, 44, 62–69, 70–83,
        84–95, 162–63, 171–72
    funding of, 76, 80, 81–82, 83, 90, 177
    Luftwaffe and, 81–83, 86–87, 94, 119,
        177, 184
    Nebel's rocketry club in, 50–51, 52–53,
        54–55, 62, 65–68, 171
    rocket-propelled fighter aircraft in, 81,
        94, 177
    secrecy of, 78–79, 80
    WVB as public face of, 72–73, 86–87
    WVB's first association with, 67–69, 71
    see also Kummersdorf proving grounds,
        rocket program of; Peenemünde;
        V-2 rockets
rockets:
    aerodynamics of, 91, 92, 99–100, 103,
        134
    alcohol and liquid oxygen propellants
        for, 77–78, 79
    arrow stability during flight of, 66, 91
    black-powder-fueled, 177
    body of, 91, 107
    engines of, 83, 92–93, 103, 105, 107, 118
    exhaust jet of, 91, 99
    explosive payload of, 90, 112
    flywheels on, 78, 79, 91
    gyroscopic control system of, 90–91,
        99–100, 103
    liquid-fueled, 43, 69, 71, 77–80, 98,
        165–66, 177
    long-range, 50, 72, 80, 90, 110
    "nose-drive," 61–62, 170
    range of, 66, 134
    recovery parachutes of, 95, 99
    solid-fuel, 41–42
    sounding, 43, 79–80, 90

rockets (*continued*)
  steering vanes of, 91, 99
  tail fins of, 91, 95, 98
  as term, 1–2
Roentgen, Wilhelm, 12
Röhm, Ernst, 63, 77
Roosevelt, Franklin D., 5, 130, 184
Rudolph, Arthur, 79, 81–82, 90, 103, 104,
    116, 126 ,166, 176, 177, 180, 185,
    187
Russell, Bertrand, ix
Russia, Czarist, 9, 11, 159
  World War I offensive of, 18–21, 158
Russian Revolution, 25, 30, 36

SA (*Sturmabteilung*), 63, 76, 163, 174–75,
    176
  "Night of the Long Knives" purge of,
    79
Saar region, 80
Sadron, Charles, 116–17, 185, 189
Sawatzki, Albin, 126, 189
Schleicher, Kurt von, 76–77, 79, 101, 168
Schneikert, Frederick P., 129–30, 190
Schröder, Paul, 103–4, 182–83
Schumann, Erich, 176
science, 12–13, 84, 85–86, 110, 176
  moral responsibility and, ix–x, 1–2, 61,
    85, 97, 128, 156, 173, 178
science fiction, 38, 39, 40–41, 42, 44, 48, 49,
    62, 67, 91, 98, 128, 135, 149
Sears, William R., 146–47
Sering, Max, 153
Seven Years' War, 9
Shirach, Baldur von, 58, 60, 169
Siemans, 92, 103, 141
Silesia, 2, 4, 20, 148
  traditionally wealthy residents of, 60,
    170
Singing Fool, The, 59
Social Democrats, 6, 25, 28, 35, 36, 49, 57,
    59, 168
Society for Space Travel (*Verein für
    Raumschiffahrt*) (*VfR*), 44, 49, 69,
    166

"Song of the Storm Columns," 59
"Sonny Boy," 59–60
Soviet Union, 50, 138, 145, 155, 172, 174
  atomic bomb tested by, 149
  in Cold War, 38–39, 127–28, 132, 147
  German invasion of, 109, 132
  Red Army of, 131, 132, 194
  rocket program of, 36, 141, 194–95
  space program of, 151, 152, 197
  in World War II, 71, 96, 106, 113, 115,
    119, 120, 178, 187
"space medicine" experiment, 60–61
space programs, 127–28, 151, 152, 197
space stations, 134–35, 149, 150
space travel, x, xi, 58
  civilian funding of, 68, 164, 172
  as Peenemünde's ultimate goal, 85–86,
    137, 139–40, 179
  rocketry pioneers' enthusiasm for, 38,
    39–41, 42, 43–44, 45, 47, 66, 133,
    166, 171, 172
  WVB as enthusiast of, 48, 49, 67, 82,
    108, 114, 118, 135–36, 137, 139–40,
    141, 149–50, 172–73
Spanish Civil War, 89, 98, 177, 182
Sparkman, John, 149
Speer, Albert, 87, 93, 119, 121, 140, 143, 174,
    191, 192
  forced labor camps approved by, 115
  Himmler as viewed by, 75, 183
  Mittelbau-Dora tunnels visited by,
    122–23
  Nuremberg trial of, 111, 187
  at Peenemünde, 113–14, 115, 117, 120,
    126, 184, 185–86
  postwar self-reflections of, 114, 131
  V-2 missiles deprecated by, 112
  on WVB's gestapo arrest, 139–40, 193
Spengler, Oswald, 37, 114
Spiekeroog, 51, 79, 167
SS (*Schutzstaffeln*), xiii, 6, 59, 71, 73–75, 79,
    129, 140, 143, 173, 174, 175, 182, 192
  army vs., 75, 102
  atrocities of, 111, 124–25, 132–33, 174
  forced labor factories controlled by, 111,

114–17, 120–21, 122, 124–25, 132–33, 185

Reitersturms (horseback riding units) of, 73–74, 173–74

V-2 program controlled by, 111–13, 114–19, 120–26, 189

WVB's membership in, 73–76, 87–88, 96, 97, 109, 111, 117, 118–19, 138–41, 173, 174, 181, 184

Stahlhelm, 35, 41, 44, 50, 66, 163

Stalag IX-C, 121–22

Stalin, Joseph, 71, 184

Stein, Gertrude, 61, 162, 173

Stettin, 29

Stimson, Henry, 137

stock market crash of 1929, 42, 43

Strauss, Richard, 89

Sudetenland, 106

"Survey of Development of Liquid Rockets in Germany and Their Future Prospects" (W. von Braun), 133–36

Swiss Institute of Technology, 60

Technik und Kultur, 36–37, 85, 114, 164

technological romanticism, 33–34, 37, 51, 114

Teutonic Knights, 8, 15, 20, 29, 173

Thiel, Walter, 92–93, 103, 185

Todt, Fritz, 184

Toftoy, Holger, 150

Triple Alliance, 11

Troost, Ludwig, 87

Truman, Harry, 130, 147, 149

"turnip winter," 25, 160

UFA (*Universum Film A.G.*), 6, 42–43, 46, 55, 98, 166

United States, 57, 71, 86, 113, 115, 126, 127–44, 145–52, 162, 164, 169, 173, 181

anti-Communist sentiment in, 38

aviation industry in, 147–48

German scientists brought to, 89, 128–44, 180, 183, 190–91

Goddard's rocketry experiments in, 36, 62, 171

guided missile production in, 150–51

Kristallnacht protested by, 107

political statements in, 5, 154

protests in 146–47

rocket program of, 36, 137–38, 144, 145–46, 148–52

space program of, 127–28, 151, 152, 197

V-2 components salvaged by, 137–38, 142, 144, 145–46

War Department of, 95, 137, 142, 143, 146

WVB feared by German colleagues in, 89, 180, 191

WVB's celebrity persona in, 88, 127, 128, 130, 135, 141, 149–52

WVB's initial employment contract in, 143, 195

Untergang des Abendlandes, Der (Spengler), 37

Usedom, 74, 83, 87, 92, 95, 111

V-1 "buzz bomb" cruise missile, 173, 185–86

V-2 rockets, x, 109–26, 127, 128, 133, 135, 136, 141, 148, 191

as contributing to German defeat, 72, 86, 173

explosive payloads of, 112–13, 119, 146

Great Britain attacked by, 130, 142–43, 190

Hitler's envisioned annihilating effect of, 119–20

hypothetical winged, 134

launch sites of, 72, 134

as long-range ballistic missiles, 72, 90

long-range heavy bombers vs., 112

military value of, 72, 76, 86, 94, 108, 112

planned fiftieth anniversary celebration of, xii

production of, 70, 72, 110–13, 117, 119–20, 126, 146, 192

production targets of, 119, 183

V-2 rockets (*continued*)
  salvaged components of, 137–38, 142,
    144, 145–46
  spiritual impact of, 113–14
  SS control of, 111–13, 114–19, 120–26,
    189
  total production output of, 190
  in U.S. rocket program, 137–38
  as *Vergeltunswaffe* (vengeance weapon),
    70, 173
  Washington, DC, display of, 145
  see also Mittelbau-Dora slave labor
    camp; Peenemünde
Valier, Max, 40–42, 44, 48, 51, 78, 165–66,
  167, 171
Van Allen, James, 146
Vatican, 155
Verein für Raumschiffahrt (VfR) (Society
  for Space Travel), 44, 49, 69, 166
Verne, Jules, 39, 40, 134, 170, 192
Versailles, Treaty of, 34–36, 50, 67–68, 89,
  163
  Czechoslovakia and, 106
  Hitler's abrogation of, 76, 80
  terms of, 34, 57, 69, 80, 169, 174, 177, 180,
    182
Victoria, Queen of England, 157
von Braun, Emmy Melitta Cecile von
    Quistorp, 14–15, 17–19, 27, 30, 36,
    77, 81, 148, 157
  appearance of, 159
  family estate of, 27
  grain reserves estimates and, 22, 159
  wartime mission of, 21, 159
von Braun, Magnus (WVB's brother), 30,
    32, 36, 128–29, 131, 141, 142,
    189–90
von Braun, Magnus Alexander Maximil-
    ian Freiherr (WVB's father), 4, 7,
    8–16, 17–31, 46, 47, 54–55, 148,
    166–67
  Berlin home addresses of, 59–60,
    169
  birth of, 8

  as Delbruck's secretary, 13–15, 22–24
  democracy opposed by, 12, 14, 25,
    64–65, 157, 171, 157, 171
  education of, 10–11
  English sojourn of, 11–12
  family estate of, 8–10, 18, 20
  family icons remembered by, 9
  gentleman as defined by, 11–12
  grain reserves estimates and, 21–22
  Gumbinnen assignment of, 29, 30–31
  Hitler and, 77, 175
  hunting by, 15–16
  Kapp putsch associated with, 30–31, 36,
    161–62
  as minister of agriculture, 49, 64–65, 66,
    68, 76–77
  monarchism of, 12, 14, 22, 36
  Oberwiesenthal estate purchased by, 60,
    175
  outpost assignments of, 27
  as parent, 16, 157
  Posen Landrat assignment of, 15–16,
    18–19, 21–22
  press office created by, 23–25, 26–27
  Raiffeisen credit union directorship of,
    31, 36, 163
  sheep and shepherd metaphor of, 11,
    156
  social milieu of, 9–10, 12, 13–15, 23, 31,
    77
  Stettin assignment of, 29
  youth of, 9, 156
von Braun, Maria von Quistorp, 148, 150
von Braun, Maximilian Freiherr (WVB's
    grandfather), 9, 10, 18, 20
von Braun, Sigismund, 15, 19, 21, 22, *32,*
    36
von Braun, Wernher Magnus Maximilian
    Freiherr:
  academic performance of, 48–49,
    51–52
  airplanes flown by, 82, 86–87, 178–79,
    187
  anti-Communist sentiments of, 138

appearance of, 3–4, 30, 55, 72–73, 86–87, 93–94, 154, 180
aviation pioneers compared to, 38, 164
birth of, 16
boyhood pyrotechnic experiment of, 49, 168
broken arm of, 131, 190
in cartoons, xii, 179
children of, 149
cosmeticized reputation of, x–xii, 38–39, 111, 169, 184
death of, xi, 128
doctoral dissertation of, 69, 78, 92, 103, 176
education of, 11, 30, 36, 48–49, 51–52, 58, 60, 69, 77, 175, 176
Einstein's letter to, 60
Frau im Mond recalled by, 48
genealogy of, 3, 4, 153
gestapo arrests of, 78–79, 138–41, 193, 194
glider piloting by, 60, 66, 86, 169, 178, 179
"Heil Hitler" as letter closings of, 81, 117
marriage application filed by, 117–18, 186
on moral responsibility, x
name of, 3, 16, 66, 72
photographs of, *32,* 53–54, 73, 76, 87–88, 118, 131, 180, 184
practical skills lacked by, 47, 52, 68, 78, 167
professional limitations of, 88
"Professor" title awarded to, 119, 133
racist policies as viewed by, 94–95, 181
remorse lacked by, 123, 131, 136–37
satirization of, 128, 189
science-fiction novel written by, 149, 197
shop floor apprenticeship of, 52, 68
as "Sonny Boy," 59–60
"space medicine" experiment of, 60–61
surrender to U.S. Army by, 128–37, 190

swastika lapel pin of, 85, 96
as visionary spaceflight hero, 38, 96, 135–36, 162–63, 181
World War I and, 19, 21
von Braun, Wilhelm, 9

"war socialism," 23
Wegener, Peter P., 87, 88, 92, 179
Wehrmacht, 63, 71, 150
Aryan descent required for service in, 94
attempted coup d'etat assassination by officers of, 107, 111
Austrian annexation and, 105–6
Hitler's control of, 101, 102, 105–6
military etiquette of, 81–82, 177
political affiliations regulation of, 97
Werewolf, 35
Wessel, Horst, 7, 155
whiggism, 38, 164, 167
Wilde, Oscar, 4, 154
Wilhelm II, Kaiser of Germany, 7, 12, 21, 26, 34, 64, 178
abdication of, 28, 161
personality of, 14, 25, 157
Wilson, Woodrow, 35, 162, 163
Wirsitz, 15, 18, 20–21, 22, 27
Wolf, Max, 165
Wolff, Theodor, 26, 27
Woman in the Moon, see *Frau im Mond*
World War I, 3, 5, 12, 17–28, 30, 33–34, 36, 42, 56, 67, 69, 81, 84–85, 94, 97, 101, 105, 114, 155, 157, 164, 171, 177
armistice of, 28
artillery weapons of, 19, 20, 28, 34, 35, 86
battle of Gumbinnen in, 19–20
battle of Verdun in 24, 160
carnage of, 18, 19, 20, 21, 33, 160
German casualties of, 19, 25, 33
as "the holy moment," 18, 158
military censorship of, in, 24–25, 28
mobilization for, 18, 158
Paris Gun of, 28, 76, 174